People and education in the Third World

W. T. S. Gould

Copublished in the United States with
John Wiley & Sons, Inc., New York

Longman Scientific & Technical,
Longman Group UK Limited,
Longman House, Burnt Mill, Harlow,
Essex CM20 2JE, England
and Associated Companies throughout the world.

Copublished in the United States with
John Wiley & Sons, Inc., 605 Third Avenue, New York, NY 10158

First published 1993

ISBN 0–582–00560–4

British Library Cataloguing in Publication Data
A CIP record for this book is available from the British Library

Library of Congress Cataloging-in-Publication Data
A CIP record for this book is available from the Library of Congress

Set in 10/11pt Plantin by 8 Q
Printed in Malaysia

To Jill

Contents

Preface

This book has had a relatively long gestation period after a fairly non-conventional conception. Unlike most textbooks, it has not developed out of any one particular course, but has grown from contributions in the last few years to a range of undergraduate and postgraduate courses in geography, education, public administration, development studies and population studies. It therefore tackles a range of issues at the various points of interaction between these disciplinary areas – issues that have been recurring points of focus and controversy. However, these are elaborated with the benefit of an even longer gestation based on experience in teaching, in development research and practice over more than 25 years, with a recurring theme of education and its role in Third World development.

I have been enormously fortunate in being able to pursue my interest in these issues in stimulating environments with supportive, exciting and convivial colleagues. Members of the Department of Geography in the University of Liverpool, both staff and students, have provided a firm base to explore an excitement and curiosity about the problems and potentials of Third World development, and successive Heads of Department have been enormously encouraging when it came to requesting yet another period in the field to consider these pursuits at first hand. In particular, I owe a great debt to the enthusiastic support and warm friendship of Professor Mansell Prothero. The interdisciplinary focus of the work has been enormously strengthened by interaction with colleagues in cognate departments, especially with those who have been associated with the Centre of African Studies in Liverpool.

In addition, however, my interests in these themes have been taken beyond the 'academic' concerns of a university environment to the 'practical' concerns of development agencies, though, as the themes in the text reveal, the distinction between 'academic' and 'practical' in development studies is not one that I have been particularly constrained by or could ever want to justify. I have been privileged to contribute to the activities of several agencies active in educational development, and to work with, amongst others, Jacques Hallak, Ta Ngoc Chau and Françoise Caillods of the International Institute for Educational Planning, Jack Maas and Stephen Heyneman of the World Bank and Jeremy Greenland of the Aga Khan Foundation. They will recognize their contributions to these ideas, and I hope will agree that working with a 'mere geographer' has been for them as valuable an experience as it has been for a geographer to have worked with them.

In the process of final preparation of the text, sections have been read and helpfully commented on by Professor Bob Woods and by Professor Dennis Dwyer, the series editor. Most of the words were processed by Joan Bell, Janet Hurst and Mary Whearty, and Sandra Mather prepared the diagrams, all with the great skills of deciphering my scribbles and of tact and diplomacy when I wanted material by yesterday and they had commitments to other colleagues. Only those who have worked in an environment with all-round support will know the debt that authors owe to their colleagues. My thanks also to the staff at Longman who had the hard task of harrying me to meet various deadlines that ensured that the work saw the light of day.

However, support must also come from any author's family. My own family has had to put up with too many absences from family activities, both when working abroad, very often during school vacations, and also at home with confinement in the study during the writing when there was music practice to supervise or gardening to be done. My special apologies are to Elizabeth for not being at home for her birthday in three successive years! Andrew, Elizabeth and David have never allowed me to forget what education and school involve from day to day, and Jill, as a teacher, has always to temper my idealism with a solid realism from the chalkface about what an education system can and cannot do, and what schools should not be expected to do.

Acknowledgements

We are indebted to the following for permission to reproduce copyright material:

The editor, International Institute for Educational Planning for Fig. 3.4 (Sudapraset *et al.*, 1980); Pergamon Press Ltd. for Table 7.4 (MacPhee & Hassan, 1990) Copyright 1990 Pergamon Press Ltd.; Oxford University Press for Fig. 2.1 (World Bank, 1991a); The University of Chicago Press for Fig. 3.6 (Plank, 1987) © 1987 by the Comparative and International Education Society. All rights reserved; the editor, World Bank Publications Dept. for Fig. 4.2 (Jamison & Lau, 1982).

Whilst every effort has been made to trace the owners of copyright material, in a few cases this has proved impossible and we take this opportunity to offer our apologies to any copyright holders whose rights we may have unwittingly infringed.

Introduction: the scope of the text

The prominence given to education by individuals, by families, by communities, by national governments and by the international community at large leaves little room for doubting its importance in the Third World. Going to school as the prime means of acquiring an education has become almost everywhere a familiar and necessary part of growing up and playing an active part in the life of the modern state. There were certainly 'schools', institutions for the socialization and broad education of the young, in pre-modern societies, whether in Koranic schools in Islamic countries, in separate temporary villages for initiating young warriors in many traditional African societies, or in the training for senior administrative positions as mandarins in Imperial China. However, the twentieth century has seen an acceptance in most parts of the world of the school as a formal institution for acquiring specific and normally universally valuable skills: at base, functional literacy and numeracy. 'Schooling' and 'education' have become almost synonymous in practice, but of course 'education' is in reality a much broader, lifelong process. The unprecedented global expansion of schools and schooling has reached its height within the last 30 years, to the extent that in the last decade of the twentieth century the majority of children in the Third World now attend school for at least some time.

The demand for schooling is strong at all levels. For individuals and their families it is a key, perhaps the key, to self-improvement and family advancement. From the point of view of villages or local communities it will allow young people to bring the benefits of education, directly or indirectly through income and influence, to the community as a whole. For governments it is seen as a necessary component of development: an educated population will more readily promote national developmental objectives. For the international community it will not only satisfy a basic and universal human need, as defined by the United Nations, but will be an investment in world development. Whether education is seen in largely Rostovian terms as one of the necessary preconditions for take-off, or in more locally defined terms as a necessary condition for self-sufficiency and individual dignity, will affect the nature and structure of the education system. The strength of demand for schooling is felt in almost all ideological and economic contexts: in relatively open, market economies like Brazil, Ivory Coast and Papua New Guinea, as well as in more tightly controlled economies in socialist or formerly socialist states like Cuba, Mozambique and Vietnam.

The universal demand has been met by major expansions of school systems in all countries. In some cases the strength of private and community demand has been sufficient for communities themselves to build and finance their own schools, as in the *harambee* self-help movement in Kenya. More commonly, however, governments have been the chief agency, with large and often growing allocations of expenditure to education to permit enrolment expansion and, where possible, qualitative improvements at all levels. In this they have been greatly assisted by financial and technical support from the international community, through the United Nations and its various agencies, notably UNESCO, and through funding by multinational and bilateral funding agencies, notably the World Bank. Government effort has often been insufficient, in quality as well as quantity, to satisfy the public demand, and in these circumstances private schools have become important. However, investment in education and schooling, great as it has been from whatever source, has seldom been sufficient to match demand, especially in countries with rapid population growth.

The education revolution has been important by any standards for most Third World countries, and the range of theoretical and practical issues arising out of it are many and varied. Those issues concerned with the planning and management of the schools system itself, including the structure of the educational hierarchy, the curriculum, the quality of instruction, the methods of assessment and the design of buildings, are *internal* aspects. Those concerned with the wider relationships of education to economic and social development, its role in promoting rural and urban change, its role in fertility decline, agricultural innovation or technological improvement, amongst others, are *external* aspects. This book is primarily but not exclusively concerned with the latter: the external aspects of education. It examines the contribution of education to several features of the development process. It does not therefore take a pre-Copernican view of education as the centre of a constellation round which other elements in the development process revolve; nor is it its purpose to review or supplement the large and growing body of literature written by education specialists that examines schools systems and their development in the Third World. Rather this book seeks to take people as its central concern, and to examine how people of the Third World, individually and collectively, have been affected by that school experience in their efforts to achieve improved economic and social well-being. Furthermore, it does so from the standpoint of a geographer with a particular concern for *development studies*.

Development studies and education

The main thrust of development studies has been to examine the processes affecting change in Third World countries, and the growth of education and the various roles ascribed to it clearly require the relationship between education and development to be a central concern. This would seem to be equally valid whatever the conceptualization of 'development'. In each of Mabogunje's four conceptualizations of development – as economic growth, as modernization, as redistributive justice, as socio-economic transformation (Mabogunje 1989) – education has a role to play. It is seen by governments and individuals as a key vehicle to increase national and individual incomes and bring *economic growth*. It is a major force for social change and *modernization* of attitudes, values and economic and social behaviour. Provision of schools and better access to education at all levels is a large and common feature of development strategies that seek to promote *redistributive justice*. The nature and structure of education can affect the speed and impact of transitions through feudalism, capitalism and imperialism to socialism in *socio-economic transformation*.

A fundamental controversy in development studies is whether and to what extent Third World countries are developing or are being underdeveloped. Are they 'developing' – becoming richer, more productive and more self-sufficient, with less poverty and unemployment, and more opportunities for individual and community improvement, and a rising quality of life? Or are they being actively 'underdeveloped' as a result of their participation in a world system dominated by Western capitalism through which there is unequal exchange of goods and services to the benefit of the richer countries, and into which the artefacts of capitalism and colonialism, including schools and structures and values creating the demand for formal education, have been introduced and which help to sustain that unequal relationship? Discussion from the perspective of this book must therefore focus on the ways in which education can be seen to affect the conflicting processes of development and underdevelopment. Under what conditions and in which respects might education be seen to promote economic activity and beneficial change for the populations of the Third World? Under what conditions and in which respects is it to be seen as an externally conceived phenomenon, based on an inappropriate Western model, for maintaining and reinforcing national and international inequalities and systems of dominance and hierarchical control?

Geography and education

The particular approach to these questions that is developed here is that of a geographer. With some major exceptions geographers have not in the past applied their specialist techniques and perspectives with any great vigour or rigour to the analysis of education systems. Most geographers have been employed as teachers or researchers within these systems, but have looked beyond them in their specific research concerns. Amongst the major exceptions might be included Liz Bondi in the the United Kingdom and Goran Hoppe and Lars Wählin in Sweden. Bondi's work has been concerned with contemporary issues in British education, notably problems of access and catchment area delimitation (Bondi and Matthews 1988). Hoppe has considered education from the perspective of a historical geographer with a careful analysis of school records in eighteenth- and nineteenth-century Sweden (Gerger and Hoppe 1982). Wählin's work is also historical in part but set in a Third World context, examining historical and current patterns of school provision and school leavers in Syria and Jordan (Diab and Wählin 1983; Wählin 1982).

In their concerns for Third World development geographers have preferred to deal with processes of rural and/or urban *production* and their physical manifestations (agricultural patterns, towns, transportation systems, migrations, etc.) rather than with the characteristics of the people of the areas studied (including their educational characteristics) or with the system of public and private *consumption*. Housing is a major and fairly well-researched example of the geographer's interest in consumption (e.g. Gilbert and Ward 1985), however, and the education system provides a further but seriously under-researched example. If geographers' concern for housing extends to the spatial, social and economic processes governing the allocation of and access to accommodation, then the geographers' concern in education must extend to the spatial, social and economic processes governing the allocation of and access to schools. Furthermore, as medical geography in the Third World has derived renewed stimulus from the study of health care provision (Phillips 1990), so the study of education provision and its consumption and associated effects provide a proper focus for a geographical analysis.

There is now a vast literature on education in the Third World that is specific to one particular country – each country has its 'Education in . . . .' volume or, more usually, volumes. Increasingly, however, education is also being considered in comparative perspective and in an interdisciplinary context into which geography and geographers have seldom strayed. Many studies by workers from other traditions, including sociology, economics, political science and education itself, have dealt with traditional geographical questions such as spatial patterns of provision and access. A review of the major journals in the field, e.g. *Comparative Education*, *Comparative Education Review*, *International Journal of Educational Development*, would show the absence of work formally recognized as geographical, but as the references in this book illustrate much research reported in these journals contributes to the debates of interest to geographers.

Even if it were possible or desirable to develop a cohesive and consistent case for a geography of education (neither of which is immediately apparent), this is not the place to do it. Rather the text seeks to explore a series of interrelated themes on the relationships between education and development in the Third World and their impacts on its people, in which the perspective and techniques of geographical analysis significantly contribute to the whole. In particular, the analysis of processes of local, regional and international integration and differentiation and a concern for regional distinctiveness, both of which are prominent in contemporary geography, are related to major themes in the relationship between education and development. Taken as a whole this approach therefore offers an interdisciplinary view that addresses concerns of a range of interests in economics, sociology, political science, demography, as well as in geography and education, and places them in the integrated perspective that is such a valuable and exciting strength of development studies.

People and education in the Third World

The first four chapters of the book consider demand for and supply of schooling and the relationships between them. Demand is considered at various spatial and social scales and from both economic and political perspectives. Why has the recent upsurge in demand arisen and how is it related to changing economic and political circumstances? How have governments been able to channel popular demand for education, and what have been their ideologies of education to create schools systems that they consider appropriate to their development strategies? The discussion of demand, focused in Chapter 1, will inevitably introduce wider issues about the perceived functions of education that recur throughout the whole text: as a force for individual opportunity and aggregate advance on the one hand, as liberal theorists and politicians argue; or as a force for repression and underdevelopment on the other, as the de-schoolers and other political radicals argue. This controversy is carried forward into Chapters 2, 3 and 4 on the supply response at the global, national and local scales respectively. The statistical description of patterns and types of enrolment growth provides an essential basis for discussion of the objectives of those institutions that have promoted and managed enrolment growth at each scale. At the global scale it examines the assumptions of funding agencies, including the World Bank and individual donor governments, and of technical assistance agencies, such as UNESCO. At the national scale it allows discussion of the range of policies of individual governments, both for quantitative and qualitative aspects of the expansions and, more specifically, their approach to regional and rural–urban imbalances in provision. At the local scale it focuses on problems of physical access to school and school–community relations, and on school location planning as a flexible set of planning techniques that incorporate demographic and locational analysis to promote simultaneously both equality and efficiency in the education system.

Some of the major effects of the expansion in enrolments and improved access to schools are the concern of Chapters 5, 6 and 7. Children in the Third World have distinctive patterns of economic and social behaviour: they are usually expected to provide domestic or even wage labour, which can seriously affect their attitudes to and achievements in school; they may also have obligations to community and family activities. After school they are normally expected to be major contributors to the household and family economy as a result of the well-paid job that their education might bring. Since these affect the demand for children, the discussion first turns to a demographic perspective on fertility and the demand for children by parents, for, as many recent demographic surveys have confirmed, there is a close relationship between education and fertility in Third World countries. Education also affects mortality and life expectancies, and these too are considered in the broader context of the role of education in the demographic transition.

Chapter 6 adopts a more directly instrumentalist perspective on the economic role of education. In particular it examines the concept of 'human resource development (HRD)' and the relationships between 'education' and 'skills', and how both education and skills, however related, contribute in different ways to rural and urban development. Chapter 7 takes this further by linking the demographic and economic aspects in highlighting the mobility of the population resource. It examines how and why migrants, both internal and international, are selected by educational achievement to the differential benefit of source and destination of the migrants.

Taken as a whole the evidence that is presented suggests that rapid expansions of schools systems have proved to be a mixed blessing for the Third World countries. In the earlier phases of the expansion they provided an opportunity for economic and social mobility and for significant improvement in economic performance. However, public enthusiasm for education has seemed to wane in many countries as national economic performance and the associated creation of new jobs in urban areas have been sluggish at best, and in many countries thoroughly disappointing. Raising levels of enrolment continues to be a policy objective in the majority of the poorest countries, especially in Africa where the effort needed is compounded by high rates of population growth. Elsewhere, notably in the newly industrializing countries (NICs) in East Asia, attention is directed more to improving quality and linking education and skill training more purposefully. Formal schooling is increasingly seen to be a necessary feature of the process of the reproduction and intensification of the internal structures of national economies that facilitates more beneficial integration into the world economic system. Although much attention has been given to alternative models of the schools system and education generally, the prospects for new perspectives on the form and function of education do not seem great, as private demand for established forms of schooling continues to be strong. The schools system is shown to reflect rather than to create the wider structures of the economy and society.

Throughout the text the term, 'Third World', is used as the primary collective generalization for the poor countries of the world. The use of the term is open to criticism, a criticism that has been intensified by the sudden collapse of the 'Second World' of socialist states from 1989 and their incipient integration into the global economic structures dominated by the 'First World' of Western Europe, North America, Australasia and Japan. Within the Third World, as normally conceived, there is widening diversity. Those countries that seem to be 'developing', in the sense of experiencing increases in per capita income levels and economic restructuring, most obviously the NICs of East Asia and Latin America, are sharply contrasted with the poorest countries of the world, mostly in South Asia and Africa South of the Sahara, where per capita income growth has been slow or has even declined, and economies are largely stagnant. That these are styled 'low income developing countries', in contrast to the 'middle income developing countries' in the classification of the World Bank (1991a), is in some sense tautological. The countries that are being considered here are largely in the 'South' in the dichotomous classification of North/South associated with the Brandt Commission (1980) and subsequent derivatives such as the Report of the South Commission, chaired by Julius Nyerere (1990). To a geographer this generalization is too gross to be consistently sustainable.

In using the term, the Third World, consideration is directed to six major world cultural realms: South and Central America, including the Caribbean; North Africa and the Middle East; Africa South of the Sahara; South Asia; East and South-East Asia; and the islands of the Pacific (to exclude Australia, New Zealand and Hawaii). Their cohesion in the face of obvious diversity and heterogeneity derives from their role as semi-periphery or periphery in the world system of capitalist production. For some countries that involvement, and increasing involvement through manufacturing exports, seems to be a major engine of development; for others that involvement seems to bring stagnation, underdevelopment and an increasing cycle of indebtedness. For almost all of them, however, some integration with the world economy and, associated with that, with broader social and cultural features of the Western world is central to the development strategy of each state. The extent of that integration is enormously variable: Saudi Arabia, for example, seeks economic integration while keeping a cultural distance; the East Asian NICs are more open to social change as an apparently inevitable by-product of the process of their rapid economic restructuring; other countries, such as China, Tanzania, Cuba and Vietnam, have sought to aspire to a socialist transition that of necessity has inhibited their integration into the world economy, but has not prevented it completely.

What they all have in common, however, is that they have sought to develop a formal system of schooling for the education of their people, following an essentially Western model, even if it takes many different forms. In this sense there is an implicit direct relationship between the First, Second and Third Worlds that sees the experience of the apparent positive relationship between education and development in the First and Second Worlds as relevant to the needs of Third World countries and their peoples in following their chosen paths to economic and social change.

A personal perspective

It is appropriate to conclude this introduction by placing it in the context of my own involvement in the issues of education and development raised in the text. That involvement began, as the careers of many students of development have done, at the sharp end of education, in the classroom teaching in a secondary school, in this case in Uganda in 1966, at a time of great optimism in that country, as elsewhere in the Third World, that education was a critical contributor to the way forward out of colonialism, underdevelopment and poverty. As a geographer I was aware particularly of the differential spatial impact of education and of large variations in access to and progress in formal schooling. Opportunity to pursue research on this theme came with appointment in 1970 to a Research Fellowship in the Department of Geography, Liverpool University, on a project directed by Professor Mansell Prothero on 'Population mobility in Africa'. Amongst other themes in this wide ranging project, I developed a concern for local mobility generated by social facilities including schools – how far were children walking to school? A study in 1971 of the journey to school in Ankole District, Uganda, was further developed by being integrated into a wider comparative project on 'school mapping' of the International Institute for Educational Planning (IIEP), an Institute of UNESCO charged with training and research in educational planning. This is an uncomfortable term for a geographer, a direct translation from the French, *la carte scolaire*, but it allowed my explicitly geographical approach to the problems of school provision to be systematically integrated into a wider set of educational themes and issues in that planning and research project (Gould 1973).

In particular, the mid-1970s was the heyday of the basic needs approach to development, led from the World Bank under its then President, Robert McNamara. Provision of facilities and opportunities for the rural poor were easily related to the micro-scale mobility studies of the IIEP project, and, beginning in 1976 in Burundi and 1977 in Haiti, then on a 12-month secondment, October 1977–September 1978, I was a consultant to the Education Department of the World Bank, developing a set of technical guidelines on 'school location planning' (a preferable but still imperfect alternative term to 'school mapping') and working on project preparation, appraisal and evaluation in several countries in the Pacific, Asia and Africa. Subsequent assignments in Papua New Guinea, Indonesia, Botswana, Zimbabwe and Pakistan on behalf of a range of agencies between 1981 and 1991, in actual planning and in training in techniques of school location planning, illustrated the widespread applicability of the geographical contribution to educational planning in rapidly expanding schools systems where equity considerations were uppermost amongst development objectives (Gould 1978). By the late 1980s, however, 'equity' was no longer such a prominent objective in development strategies and schools systems were no longer extending their spatial coverage. Different issues, such as privatization and decentralization, have come to the fore to attract the geographer's attention, and there are currently several unresolved and contentious issues in educational planning, such as decentralization and local variation, where a geographical perspective can make a substantial analytical impact (Gould 1988b).

These contributions to planning and implementation have been of value in training programmes for educational planners in increasingly decentralized education structures. With large expansions in the numbers of schools and pupils new systems of planning and management were developed, and particularly at the local level, with training needed in statistics, planning and budgeting and facilities management. IIEP in particular, as the UNESCO agency responsible for training in educational planning, has taken the lead in developing appropriate training strategies and materials in Africa, at a difficult period in its economic progress when it is suffering the financial problems associated with indebtedness and 'structural adjustment' programmes imposed by the IMF or the World Bank. I have been able to contribute to those training activities in Zimbabwe and Botswana, and most recently in Ghana since 1988 in developing technical guidelines and contributing to several training courses to elaborate them, mounted as part of a fundamental decentralization of the organization and management of its Ministry of Education (Gould 1990a).

These several direct involvements with governments and with international agencies in the workings of education systems have to be set beside a broader baseline commitment to teaching and research from 1974, working in a geography department and specializing in population and Third World development topics (Dickenson *et al.* 1983). Research in population migration in particular, supported by grants from the then Social Science Research Council of the United Kingdom and the British Academy, allowed the fostering of a particular link between education and migration in the development of a tracer study (still continuing) of school leavers in Kenya. This has subsequently been extended in a parallel study in Pakistan supported by the Aga Khan Foundation. At a much higher level of educational achievement, I have been able to link this work with studies of the international migration of skilled and educated workers in Africa (Gould 1988c). The population-based research was informed by and indeed contributed to the directly educational projects.

This book links practical experience in educational development in a wide range of Third World countries with an academic concern for the ways in which people and governments in Third World countries use their own internal and external resources to provide a better quality of life for their citizens. Though it is an exploration in what might loosely be described as applied geography, the text offers few prescriptions. However, if Julian Simon (1981) is right, and I believe he is, and population is indeed 'the ultimate resource', then its mobilization through education must constitute a vital ingredient in our appreciation of the processes of development.

Chapter 1

The private and public demand for education

Early on weekday mornings, often at lunchtimes, and again in the evening, the roads of most towns, villages and rural areas of the Third World are thronged with children going to or coming from school. They are happy, boisterous and noisy, shouting, laughing and playing, often in well-kept and neatly-pressed school uniforms that can seem at odds with the surrounding poverty of their families and villages. Their familiar presence is an indicator not only of the high proportion of the population in the school-age groups in most Third World countries, but of the massive presence and importance of schools and schooling. Schools are generally the most widespread of the services provided by government. There are, for example, usually far more schools than health posts, and they are more likely to be found in small communities and in the remotest settlements. Recent expansions of schools systems have been rapid, and there are still some children who for reasons of physical access or social and economic constraints are denied a place in school, but the majority of children in the Third World can now attend school for at least some of their school-age years. Demand is everywhere high and the popular thirst for education seems to assume enhanced life chances in the future to the children themselves and to their parents and communities.

At the same time the governments of all Third World countries affirm the need for an educated population as necessary for the modern state, one of the preconditions for development. Development, however defined, will be more rapid and better appreciated with a more educated population. Though the detailed nature of educational needs and priorities differs very greatly from one county to another – some countries emphasize primary schooling, others higher education; some countries emphasize general education, others vocational and practical subjects – these are differences of emphasis and detail rather than principle. In order to implement that principle, governments have invested large proportions of an often meagre government budget in education.

Both the private and public demand for schooling are great, and each reinforces the other. In this chapter the main features of public and private demand are explored, identifying their historical and contemporary complementarities, for it is impossible to identify separately the one as having precedence over the other. Did the 'chicken' of private demand come before the 'egg' of public demand, or was it created by public demand? In the final section of the chapter voices of scepticism are identified. Despite the obvious public and private demand and the linkages between them, education as provided in the Third World may not serve the expected needs of the individual or of the state. Has the burgeoning demand been built on false premisses and unrealizable expectations?

Private demand

The rapid growth of enrolments at all levels of education in almost all Third World countries in the last 30 years suggests that that popular thirst for schooling is global, part of a universal movement and associated with wider aspects of the development process. Formal education was largely a creation of the colonizing powers, transferring the institutions for educating and socializing young people, i.e. schools, into the colonies or dependencies. In Europe and North America public financial commitment to schooling was mostly established only in the nineteenth century, and its transfer to the colonies was entirely consistent with the broader 'civilizing' objectives of colonialism in that period and into the twentieth century. The transfer may have been managed directly by the state or indirectly through mission churches which were closely associated with the colonizing state (Holmes 1967), but the net effect was roughly similar.

The colonial presence created a class of educated indigenous citizens who became the reference models for many. The *evolué* of the French colonies, the man (or, less commonly, woman) who through education had acquired French language and culture, could find a favoured place in the economic and social life of the colony or even in France itself. The respectful, educated professional or civil servant of the British colonies, as personified by E. M. Foster's Dr Aziz of *A Passage to India*, or a figure of fun like Joyce Cary's *Mister Johnson* in Nigeria, aspired to the colonial cultural values, and provided an implicit model for others. The educated were figures of social respectability and economic substance in comparison to the mass of colonial subjects. That respectability and substance were widely perceived, by those who had been to school as by those who had not, to be in a major sense due to their having become 'educated', i.e. attended a formal school with a recognizably European curriculum. Educational attainment was a principal criterion used for entry into the civil service or other modern sector positions, as it was only by the late nineteenth century in Europe itself. The colonizers needed a small educated workforce to maintain the administration of the colonies, and the state or missions provided that labour force with schooling that was usually similar in content and style to the liberal general education of the European middle class. In these circumstances, numeracy and literacy were much more highly prized than the acquisition of technical or practical skills (Watson 1982).

The needs of the state and the opportunity for employment by the state were not the only reasons for an early appreciation of the value of education. The requirements of the Church to learn the catechism or to read the Bible brought the Christian missions early into providing schools, very often at a very low and basic level. Still today, in some countries where schools are provided by Christian missions, there is often a distinct cleavage between a few 'better' schools where a whole cycle is provided and the so-called 'bush' schools where religious curriculum dominates. In Burundi, for example, the mass of the numerically dominant Hutu tribe disproportionately attend *yagamukama* ('The voice of the Lord') schools for basic catechistic training, while formal schools in the public sector (mostly founded by the Christian missions) since the early German and Belgian colonial periods and continuing through to independence are disproportionately attended by the politically dominant Tutsi. The structures and implications of educational provision in Burundi mirror those of the ethnically polarized state with its 'premiss of inequality' (Greenland 1974).

In Muslim societies, however, from Morocco and Mauritania in the West to Indonesia in the East, there is a clear division between the role of the school for religious education and the role of the school for modern development. The Koranic school, with learning in Arabic led by a *mallam*, is an essential part of the upbringing of a Muslim, and all children, boys and girls, are exposed to Koranic recitation, and many progress to higher religious studies. In many Muslim countries, e.g. Mauritania (Botti *et al.* 1978), there has been a systematic attempt to link secular and religious education, with *mallams* being given formal training and qualifications in teaching and with Koranic studies formally integrated into a more general curriculum, but without obvious success in improving learning outcomes. The demand for secular education, especially for girls, has been conspicuously lower in Muslim countries than elsewhere, partly because of the obvious difficulties of complementarity with the Koranic school (Bray *et al.* 1986:79–96). Nevertheless, Muslim countries have not been isolated from the global trend of rising enrolments in secular schools, and the state has often taken the lead in secularizing education, as in Egypt or Turkey. Yet in many countries the re-emergence of religious fundamentalism and growing dependence on the *Sharia* or Muslim law have been associated with a flourishing of Islamic studies within the secular curriculum. In Pakistan, for example, Islamiat is a compulsory subject in the curriculum in all state schools.

Private demand for education can vary considerably from country to country. It seems to be greatest in those countries which have experienced independence since 1950 (predominantly in Africa) where, as a result, there have been many new job opportunities, especially in a rapidly expanding bureaucracy. It is also strong in the NICs of East and South-East Asia where rapid technological progress and a rapidly expanding labour market for skilled workers fuel the private demand for education (Postlethwaite and Thomas 1980; Thomas and Postlethwaite 1983). Furthermore, incomes, both public and private, have risen substantially in those countries, and can be invested in meeting that demand. In India, despite its relative economic stability without rapid technological change, there is still a very strong private demand for schooling at all levels, one hardly affected by high unemployment rates and relatively low incomes for many educated people. In that large and complex country, education is widely perceived to be a principal requirement for individual economic and social advance. In Latin America the demand is much higher in urban areas than in rural areas in absolute terms (the urban proportion being over 60 per cent of the total population by 1990) and in relative terms, for it is in urban areas that there is greatest economic dynamism, and it is in these areas that there are sufficiently high and numerous private incomes to sustain that demand (Brock and Lawlor 1985).

Throughout the Third World an individual's level of education is a major criterion for his or her lifestyle, career opportunities and life chances generally. Those who are wealthy and have secure and well-paid careers and a high 'quality of life', by most objective measures, tend to be those with some education. Those who are poor, in low paid or insecure jobs, inadequate housing and poor health tend to be those with little education. Of course there is no perfect correlation – some very wealthy and comfortably off people have never been to school; some educated people are impoverished. However, it is clear to all that level of education is important to life chances. It is certainly not a sufficent criterion, nor is it even a necessary criterion, but the acquisition of formal qualifications is a major advantage that few would choose to be without.

That this should be so is justified by economists in terms of the high private rates of return to education. Using a cost/benefit framework, many studies in a wide range of countries, e.g. India (Heyneman 1979), Kenya (Thias and Carnoy 1972) and Brazil (Singh 1992), as well as in international comparisons (Psacharopoulos 1981; Woodall 1973), have shown that there is a strong and positive correlation between lifetime earnings and level of education. The measurable benefits of earnings outweigh the costs (in direct costs of schooling and in forgone earnings while in school) by a large margin. The rate of return on any costs incurred in school attendance over a number of years is calculated as the rate of interest on the investment needed to have an income equal to the extra income to be derived as a result of that period in school. Calculated annual rates of return to the individual for investment in primary schooling are typically over 15 per cent and may in some countries be over 50 per cent. Costs are very low at this level, and benefits are substantial. For secondary education, where direct costs are higher and forgone employment earnings are more likely, annual rates of return are lower, but are typically over 10 per cent and may in some countries rise to over 30 per cent. Psacharopoulos and Woodall (1985) provide a global review of private rates of return to education in Third World countries and conclude that education is a very highly productive private investment. It is more productive than most other investments most families can make: 'for the individual student or family education is usually a highly profitable personal investment. The expected benefits more than compensate for the burden of high costs, including earnings forgone' (Psacharopoulos and Woodall 1985:119). Their work, done within the World Bank, has justified the World Bank's previous commitment to investments in education, and is continually invoked to justify continuing commitment in the face of competing demands on its resources (e.g. World Bank 1991a).

This approach, with its strongly positive conclusions and its policy implications, is not without its critics. The calculated rates of return are based on historical earnings, and it is clear that earnings patterns in many Third World countries have changed in the 1980s under austerity and structural adjustment programmes. In particular, in many countries public service salaries have dramatically fallen behind rises in the cost of living. In geographical terms there seems, particularly in Africa, to have been a narrowing of the rural–urban income gap, due to the downward trend in urban incomes rather than an upward trend in rural incomes (Jamal and Weeks 1988). Returns to education have fallen substantially in these conditions. Furthermore, the measured rates of return are very high partly because of the absence of realistic measures of alternative opportunities, especially in agriculture. Measures of income in that sector are so notoriously incomplete, conceptually difficult and generally underestimated, that a relatively easily calculated measure like modern sector employment income looks attractive in comparison. Calculated returns to rural primary schooling in Brazil are estimated to be in the region of 14–15 per cent, lower than in many Third World settings, but still significant. However, enrolment rates in rural schools in Brazil are low, suggesting relatively high cost of schooling and a perception by the population of rates of return to primary schooling that are significantly lower than the economists' calculated rates (Singh 1992). Although the picture generally is probably not as bright as calculated crude rates of return suggest, and is likely to be deteriorating in most countries with the over-supply of educated people relative to the needs of the economy, the general argument remains sound: expenditure on education is a productive investment for most people in most circumstances.

Socially, too, education is attractive. It offers a prime mechanism for social mobility through its links to employment and job status. Education provides individuals with communication skills in a national language, often English, French or Spanish, to set them apart from those who may only speak a local or tribal language. It inculcates cultural values and behaviour that will make the individual more socially acceptable in the élite. Elites are a key feature of most Third World societies. They are characteristically small, cohesive and politically articulate, and a high level education may be necessary for entry to them. Access to the élite was relatively open where there have been massive social changes, especially since independence in Africa in the last 30 years. In these circumstances it was not uncommon for the sons and even the daughters of poor farmers to gain élite status, both in the social and political sphere, and there was considerable social mobility through education, e.g. in Ghana (Foster 1965) and in West Africa generally (Peil 1977:189–201). Many African presidents have been or are from poor backgrounds, and have usually made great political capital out of their humble origins. However, even in Africa relatively open access to élites has rapidly declined as the new élite has consolidated its position in education as in other aspects of its power and patronage. Elite families are much more able to afford to send their children to school for longer, and to the better, often private, schools, creating part of the basis for a class-based social structure, within which upward mobility is much less easy than it was previously.

Africa is moving very rapidly towards the situation that has been more familiar in Latin America where class divisions dominate all aspects of social and political life. In Latin America prospects for social mobility through education are not as great as they have been in Africa, and here too are probably declining. The gulf between urban and rural life, between the urban élite and the *campesinos* is large and exacerbated by considerable variation in the type and quality of education. Private schools are familiar in urban areas, and have in the past dominated the expansion of enrolments in most countries (Brock and Lawlor 1985). The state has often created separate rural and urban schools' directorates, differentially funded. Jean-Pierre Jallade (1974) has shown how it is the élite who benefit from subsidies in higher education, in Colombia specifically through the bursary system. Yet even there, with the balance of opportunity heavily loaded against the urban poor and rural people in general, the demand for an education that can facilitate but cannot guarantee escape from the economic struggles and social indignities of the *favella* and the rural slum is all too evident. A similar situation is found in the Caribbean. In Jamaica the demand for schooling remains high even though the education system has been shown to be a force for maintaining and reproducing the social order (Strudwick and Foster 1991).

In the political sphere education commands respect. The uneducated politician has little claim to national political office, though education is probably only one of several necessary attributes, like an identifiable class-based or ethnic or regional constituency. In many Third World countries trade union leaders are much better educated than their rank and file, and their education gives them the authority, as well as enhancing the ability to lead. The personal charisma of an Oxford-educated Benhazir Bhutto, for example, enhanced the family and historical basis of her surprising election as the first female Prime Minister in the Muslim world. Conversely, it is incomprehensible to many in the Third World that a country with a highly educated population like the United Kingdom should have in John Major a Prime Minister who does not have a university degree!

Education, then, provides the individual with economic, social and political advantages. Since it provides all three and not merely the first, expenditure on education needs to be seen to be not only *investment* but also *consumption*. Economic theory is very ambiguous on this point. In classical economics education has tended to be viewed as a consumption good, requiring investment for personal development and benefit in the short term and financed from current income. With the growth of the human resources school, after Schultz (1981) amongst others, education has been taken very much as an investment in human capital, with long-term benefits both to the individual who is educated and to the public at large. As identified above, the main argument in recent literature follows this reasoning and uses rates of return to justify the comparability of public education with public investment in other productive and infrastructural sectors of the economy. Private expenditure can be similarly justified. Paying school fees and other educational costs is, in these terms, a 'rational' use of scarce household resources, with greater long-term return to individuals and to households than expenditure in other directions, notably in agriculture.

While this approach is now generally accepted and has given great stimulus to systematic research on the economics of education, there are clearly circumstances in which expenditure on education is more closely akin to expenditure on consumption goods, like food or clothing, in that it is necessary for individual or household survival in the fairly short term. In focusing on private demand, as we have done in this section, clearly the investment/consumption question is a major issue, for educational expenditure it can clearly be both. Maragoli, Western Kenya, for example, is an area of very heavy out-migration where the majority of households are dependent for most of their cash income on the remittances of male migrants who work elsewhere in Kenya, but mostly in Nairobi. In this densely populated and impoverished area relatively few cash crops can be grown as almost all the land is needed for family subsistence production. But cash incomes are derived mainly from urban sources. Joyce Moock (1973) has described Maragoli 'as a dislocated Nairobi suburb'. The educated are more likely to find a job in the capital, and there, as throughout Kenya, there are sharply increasing income returns to each additional level of educational attainment. Typically over 20 per cent of household income in Maragoli is allocated to educational expenditure to maintain the flow of income by school leavers continuing to have access to the urban labour market. Expenditure on school fees under these conditions must be seen as consumption for household survival rather than investment for household improvement (Martin 1982).

In this Kenyan case the element of choice between alternative expenditures is severely constrained. More typically, expenditure choices are to be made, and, whether for consumption or investment, education usually ranks high among them. Expenditure on education may for the rich constitute 'conspicuous consumption', in the derogatory sense of that term, satisfying mainly social or strictly private wants or expectations. However, children and parents generally appreciate the potential benefits in a range of contexts to the extent that what is investment in one context (e.g. to raise lifetime or even short-term earning expectations) may constitute consumption in another (e.g. to provide the context for socialization of a child into the wider social and cultural norms of society). Nevertheless, the strictly functional view of educational expenditure, that it is for private benefit, does provide a general justification for arguments of the human resources school that see the substantial expenditure on schooling as a response to rational investment choice for most families in most countries of the Third World.

Public demand

Private benefits find a parallel in the public or social benefits of education: those benefits that accrue to society and the economy as a whole and not merely to those who benefit directly as a result of their own education. The public benefits of education are most obviously felt in the economic sphere. An educated population is necessary for an increasingly complex technology, with more specialized and highly trained manpower needed to undertake a widening range of increasingly sophisticated tasks. Some training can be done on-the-job, and the amount of on-the-job training needed can vary a great deal, not only from job to job, but also for any given job from country to country or even factory to factory. Yet it is clear that a basic educational background is needed as a foundation for the training. As a minimum, a modern state needs a population with a widespread basic literacy and numeracy if it is to have sufficient people to train for more specialist tasks. Many education systems have been guided in their expansion and shape by the manpower needs of the economy. The principal function of the education system in these terms is to produce an appropriate amount and mix of manpower. Government expenditure on education to achieve such a supply of trained workers is premissed on the benefits to the economy as a whole and the increased potential for modern production that it brings.

In this perspective the public benefits of education are closely linked to structural changes in the economy, and particularly with increasing industrialization and urbanization, but similar arguments may also be applied to the agricultural sector. Agriculture is more productive and farmers more likely to innovate and apply advances in technology where they have been to school. And the longer they are in school the more likely and the more rapidly the increases in productivity will come (Jamison and Lau 1982). Even for the poorest and most traditional subsistence agricultural economy, more education for farmers will be expected to bring production benefits, but these will be greatest in more modernizing economies.

The evidence for such a positive view of the role of education is by now well documented. While classical economic theory largely ignored the role of human capital in economic production, seeing labour as brawn power rather than providing brain power, the growth of the human resources school of economists brought education as investment in human capital firmly into the overall calculus (Becker 1974; Schultz 1981; World Bank 1991a:52–69). Negatively, they showed that aggregate increases in production and output, in developed as in Third World economies, could not be attributed solely to investments in physical capital, and that the residual was due to the effects of education. Positively, they calculated in a range of economies in developed and Third World countries the rates of public return on investments in education. Their arguments have been used by individual governments and international agencies, notably the World Bank, to justify public investments in education (Psacharopoulos and Woodall 1985). Since it is the public purse that provides most or even all of the costs of education, public rates of return are generally lower than private rates, but they are still sufficiently substantial to justify large investments in schooling. Investments in education are generally calculated to be more productive than investments in most other sectors, including agriculture and manufacturing. Together with investment in health, especially primary health care, education is investment in human capital, boosting the economic productivity of individuals. In Julian Simon's (1981:348) phrase: 'The ultimate resource is people', and like other resources needs to be managed. The education system is a major element in the management system that is required. The theme of 'human resource development' and the role of the education system in that development is further elaborated below in Chapter 6.

However, the public benefits do not lie solely in the economic sphere. There are also social and political benefits that justify public expenditure. These, however, are rather more controversial and more amenable to political manipulation. In many countries schooling is held to be a major force for socialization of children into the culture and ethos of the state. At all levels there may be formal classes in citizenship or public awareness in the official curriculum. More importantly there may be a hidden agenda in the curriculum as a whole, and especially in history classes, that inculcates the ethos of the state. This may be capitalist or socialist; it may be fiercely nationalist or even regionally specific, but there will normally be an attempt by government through the curriculum to foster national goals. This of course can lead to 'political' education, if not indoctrination, and clearly there is a possibility of the education system being used to seek political rather than economic objectives as a primary goal. In general, however, the education system contributes to the socialization of children, and in so doing brings positive social benefit.

That there are major public economic and social benefits arising from education is very much a view developed in the last 150 years. Before the nineteenth century and the era of mass public education in Europe, in most countries formal schooling was very much for the élite and presumed to bestow only private benefits. If there was to be public benefit from education it would arise out of apprenticeships and other on-the-job training. In a rigid pre-industrial society social mobility was certainly not encouraged by dominant groups in the state. Since education could be seen as a vehicle for promoting social mobility, it was felt by many to be positively dangerous. The educated did not know their place, and were inclined to be trouble-makers! Education could be socially disruptive and a force for change, and there is ample evidence from European history to justify such a defensive argument.

As far as the Christian Church was concerned, though, schools did have a valuable public function in propagating its teachings, but even here there was widely different experience which was reflected in Third World countries. Gunnar Myrdal (1968:1633–5) contrasts the experience of colonial Philippines with colonial Indonesia in this respect. In the Spanish Philippines, over a period of more than 200 years from the seventeenth to the nineteenth century, the Roman Catholic Church established a widespread and large schools system whose principal purpose was strictly religious, and most of the population were converted to Catholicism. The deep roots of literacy established then, and subsequently fostered by US teachers and policy in the first half of the twentieth century, have been important for contemporary high rates of literacy and numeracy and current high levels of demand for education. In Dutch East Indies, on the other hand, most people have remained Muslims (or Hindus on Bali) and the Christian missions were much less active in providing education. The colonial state, however, through the Dutch East India Company, provided schooling only for a small number of Indonesians, sufficient to satisfy the administrative demands of the Company for clerks, etc. Here the base was narrow, and until recently has remained so, as the public demand for secular education remained small.

The Indonesian experience is more typical of much of colonial Africa, where the immediate administrative needs of the colonial state dictated the pace of growth and type of education that was provided. The public demands were for clerks rather than technicians, hence a bias towards literacy and numeracy, and towards general education rather than vocational or practical education. The skills necessary for practical tasks could be acquired at the workplace and through apprenticeship schemes, formally or informally organized (Whitehead 1982). In Latin America the base was broader, and was more similar to the Philippines model, with the Church rather than the state providing a public justification for schooling (Brock 1985).

After the Second World War, however, the emphases

changed, and the public function of education became much more prominent. In the 1950s the United Nations had declared education to be a basic human right. Countries that had won independence from a colonial power inevitably sought to establish societies based on general notions of equality and human rights, and their support for a social justification of public education was unquestionable. If for no other reason, the state could justify its expenditure in education on the principles of equality of opportunity, and could design its programmes to implement these objectives along the lines set out by UNESCO regional conferences. For Asia in Karachi, Pakistan, in 1960, for Africa in Addis Ababa, Ethiopia, in 1961, for Latin America in Santiago, Chile, in 1962, and for the Arab States in Tripoli, Libya, 1966, the regional conferences set targets for enrolment expansions amongst other quantitative targets, in all cases to achieve universal primary education within a limited time horizon, mostly to 1980 (Fredriksen 1981). These very ambitious targets have seldom been achieved in practice, but in the euphoria of the period after independence for so many countries seemed to be not only reasonable but achievable.

These targets, however, were set with economic development in mind, for it was then evident that Third World development necessitated a much larger and better supply of educated people than had hitherto been available. Quantitative expansion had to be the first priority and was taken up in almost all countries, but with larger proportional increases at higher and secondary levels in the first instance. Governments were embarking on ambitious development plans which depended on a growing supply of skilled workers. In addition, in many countries, especially in Africa, there was strong political pressure to indigenize the labour force, as colonial expatriates left to be replaced by citizens, thus further increasing the need for expansion of the schools system.

In this expansion, countries were supported by bilateral and multilateral donor agencies, providing funds through loans and grants for school building programmes that would not have been available from internal revenues. Loans could be justified by the economic calculus of high rates of return as measured by standard economic accounting techniques. Furthermore, since it could be shown that returns to primary education are higher than returns to secondary education, the emphasis, particularly in the 1970s, shifted to investments in primary schooling. This too was consistent with a basic needs thrust in development strategies, encouraged particularly by the World Bank at that time. Here again, the social and economic justifications for increased public expenditures had come together.

The importance of education to the public domain necessitated the direct control by the state in controlling the nature and pace of expansion through a willing acceptance of the theories and practice of educational planning. As a systematic approach to setting educational objectives, marshalling and managing resources and implementing reform programmes to achieve these objectives, educational planning developed rapidly in the 1960s and 1970s, guided particularly by the international agencies, notably UNESCO. These agencies sought to establish a framework for appraising and justifying investment alternatives in this sector by supporting institutions, notably the International Institute for Educational Planning (IIEP), for research and training programmes to create a cadre of officials to use and develop the growing array of planning techniques in education. External pressure for systematic planning by international agencies further promoted the attention given to education by governments, and brought forth many national plans for education. These were normally integrated into and were consistent with the wider context of national planning, and were often based on reports of national education commissions which established the objectives and structures on which the planning could be based.

Such a commission in the 1970s produced for Botswana, *Education for Kagisano: Report of the National Commission on Education* (Botswana Government 1977). The Commission comprised six commissioners, one a local Member of Parliament, another the Rector of the then Federal University of Botswana, Lesotho and Swaziland. The other members, including the Swedish chairman, were internationally recognized 'experts': an Ethiopian (subsequently Director of the Education Department of the World Bank), an American and an Englishman, and reported 'to identify the major problems affecting education in Botswana and the issues of principal concern to the Government of Botswana' and 'to present recommendations regarding effective implementation of an effective programme to overcome problems and achieve goals'. The Commission was established in 1975, some nine years after the country's independence, and reported in 1977 with recommendations for structural change and expansion at all levels of the system that were accepted at the time. These remain the basis for educational planning in Botswana into the 1990s.

In Botswana educational planning has reflected the international consensus. It is well integrated into the broader ethos of planned development in both economic and social spheres as an unquestioned feature of the national development stategy. The approaches and objectives of educational planning in practice vary widely from country to country, but, given the near universal presumptions of the positive public benefits from educational expenditure, the central role of the state in directing the pattern of expansion is everywhere accepted.

Voices of dissent

Amidst the widespread private clamour for education and the support given to it by governments, there have been voices of scepticism and outright dissent from the positive views towards the private and public benefits of schools and schooling that are the basis of official planning assumptions. These voices have argued that the case for education has been not only overstated, but, much more fundamentally, also mis-stated. Criticisms have come from the political Left and from the political Right against the liberal consensus view elaborated in the earlier sections of this chapter. From the Right the main thrust of criticism, in developed countries as in the Third World, has been levelled against the state becoming over-involved in education, with implications that are not only economically unsound but politically dangerous. From the Left, also in developed as in Third World countries, is the accusation that formal education as developed in the twentieth century reflects existing structures, and principally the capitalist and neo-colonialist bases of most societies, rather than changing them, that it has been a force for social and political inertia and economic backwardness rather than change, and that claims for its leading role in economic development in Europe and North America in the past, and by implication for the Third World in the future, have been much exaggerated.

Such a negative view of the role of education in development is evident in some 'liberal' writings. In the United Kingdom, for example, studies over many years, written from a wide variety of perspectives, have emphasized the persistence of social immobility in a class-ridden society, despite the apparent liberalization associated in schools with the Butler Education Act of 1944 and the growth of comprehensive schools systems in the 1960s, and in higher education with the post-Robbins expansions of the 1960s and 1970s. However, more trenchant criticisms came from Marxist and neo-Marxist writers by the 1970s, such as Bourdieu in France, Raymond Williams in the United Kingdom, and Bowles and Gintis in the United States (Dale *et al.* 1976), arguing that schooling had become a tool for production in, and the reproduction of, the structures of capitalism. Education, it was claimed, perpetuates and institutionalizes inequality, and allocates resources for the benefit of the production process at the expense of education for individual betterment. Such a view has been confirmed, indeed strengthened, by the economic restructuring of developed countries since the 1980s. This involved an intensification of 'post-Fordist' production processes in which developments in education, such as stronger linkages between the education system and the labour market, seem to illuminate the role of schooling as a servant of the dominant capitalist order. Education has reflected rather than created or even altered the essential bases of capitalist economic and social structures.

Extending such conclusions to the Third World would lead inevitably to a view that saw the education systems as an integral mechanism of the penetration of global capitalism. Education in these terms must be closely associated with colonialism and neo-colonialism as a means of social control. In politically independent but economically and culturally dependent countries of the global periphery and semi-periphery, education and expansions of schooling have assisted the development of both indigenous capitalist production and the penetration of multinational companies. The costs of training a labour force are borne by the state and by individual consumers, but the principal beneficiaries are capitalist producers. It may be possible in theory to have schooling that does not serve the requirements of capitalism, as in the objectives of education in socialist countries, but it needs to be integrated into a wider set of structural changes in the political economy, as part of a process of socio-economic transformation as the base objective of 'development' (Groth 1987). In Cuba, for example, basic rural secondary schools were established from the 1960s explicitly to further the socialist objectives of the state by better integration of curriculum and introduction of productive work as part of rural transformation: 'That educational revolution is the result and at the same time the means of establishing and consolidating the social revolution is revealed by . . . the *Escuela en el Campo*' (Figueroa *et al.* 1974:iii; see also Richmond 1985). In China there has been a similar presumption that the schools could and should contribute to revolutionary change (Hawkins 1983). However, in the majority of Third World countries, tied to the international economy and indebted to it through commercial and official loans and 'structural adjustment policies', education systems have mostly continued to reflect the broader liberal and universalist consensus associated with economic political and cultural dependency, and have seemed strenghtened by it.

A related set of fundamental criticisms is associated with Ivan Illich and the de-schoolers. Arguing from the basis of his experience in Puerto Rico and Mexico, as well as in metropolitan United States, Illich believes that the institutions of education, the schools, and the practitioners, the teachers, have been in the past and continue to be stultifying forces that have operated a system that acts against the interests of the poor:

> Equal educational opportunities is, indeed, both a desirable and feasible goal, but to equate this with obligatory schooling is to confuse salvation with the

> Church. School has become the world religion of a modernized proletariat, and makes futile promises of salvation to the poor of a technological age. (Illich 1973:18)

For Illich the private thirst for eduction is based on a false consciousness of the effects of schooling, a consciousness promoted by the state through its control of the public media as well as by its control of the curriculum in the schools themselves. It is an obviously successful confidence trick.

While many socialist and 'liberal' observers accept many of the conclusions of Illich and others on the practice of education in the Third World and its observed effects, the majority adopt a more positive stance towards the potential of the education system simultaneously to promote appropriate economic development and to enhance the life chances of individuals who acquire a formal education. Correctly planned, education can be a strongly positive force in the national effort to raise incomes and redistribute wealth. Paulo Freire, influential in his native Brazil and elsewhere, Julius Nyerere from Tanzania, Mahatma Gandhi from India offer three examples, each coming to the problem from different directions, of positive solutions in education within a radical and egalitarian perspective.

Freire's *Pedagogy of the oppressed* (1972) recognized the role that schooling had had in Brazil and elsewhere in Latin America in maintaining the poor as an underclass. His involvement with adult literacy classes and promotion of revolutionary, liberating thought or 'conscientization' in these classes, revealed the possibilities for a restructured and de-institutionalized education system taking a leading position in social change. This view is clearly linked ideologically as well as personally with Illich and integrated with broader trends in social thought, such as liberation theology. Freire's ideas have been influential in moulding formal education systems in a few revolutionary countries, including the former Portuguese colonies in Africa, Mozambique, Angola and Guinea-Bissau where his writings were immediately accessible. They have been even more influential in non-formal education, whether provided directly by the state or by institutions like the Roman Catholic Church, promoting conscientization and a liberation theology often in conflict with the state authorities and, for individual priests, with the ecclesiastical hierarchy, raising the awareness of the poor to their own situations as a necessary basis for inducing fundamental change.

A less overtly subversive but nevertheless revolutionary approach is associated with Julius Nyerere. As President of Tanzania, he introduced *Education for self reliance* in 1967 as an integral element in a larger package of reforms at that time designed to restructure Tanzanian society along socialist principles:

> Our village life, as well as our state organisation, must be based on principles of socialism and that equality in work and return which is part of it. . . . That is what our education system has to encourage. It has to foster the social goals of living together, and working together, for the common good. . . . Our education must therefore inculcate a sense of commitment to the total community, and help the pupils to accept the values appropriate to our kind of future, and not those appropriate to our colonial past.
>
> (Tanzanian Government 1967:7)

Reforms in curriculum (e.g. compulsory agriculture and use of Swahili, the national language, throughout the whole basic cycle) and organization (e.g. decentralization of responsibilities to provinces and districts, severe restriction on cxpansion of secondary school places in favour of achieving universal primary schooling and much expanded adult education) sought to turn the theory into practice. Nyerere himself, as national President and leading ideologue, was as closely associated with the practice and implementation of the reforms as with their formulation, easily justifying the Swahili ascription, *Mwalimu*, the teacher.

In the 1920s and 1930s Gandhi was a leader in the basic education movement in India. He, with others, sought to develop a practical vocational bias in primary schools that would be more 'relevant' to the needs of the majority of Indians and to India as a whole. This bias involved training in rural crafts, such as spinning and pottery, using traditional technologies:

> the child at the age of 14, that is after finishing a seven years' course, should be discharged as an earning unit. . . . You impart education and simultaneously cut at the root of unemployment.
>
> (Gandhi 1950, quoted in Sinclair with Lillis 1980:51)

Such a view of schooling fitted well with the wider strategies for Indian development proposed by Gandhi before independence. The emphasis on *relevant* education has been a now familiar positive response to these more fundamental voices of dissent, a theme that is as apparent in the more recent approach of Nyerere as it was in the British colonies generally in the 1920s (Gould 1989; King 1971).

These are three examples, each from different continents, of education and schooling being placed by revolutionary figures in the forefront of political change. The examples do not lend support to the view that education can do no more than exercise a negative and essentially anti-progressive influence on economy and society, but that it can be a positive force for change. Though, as discussion in later chapters will argue, many of the attempts to induce social change through education have proved to have been, at best, disappointing in practice, the Left does recognize the revolutionary potential of education in the appropriate conditions.

At the opposite end of the political spectrum, the criticisms from the Right of the expansion of public education are also substantial. These have been voiced over a longer period, stretching back to Adam Smith and the very basis of classical economics. There are basically two types of criticisms, each related to the other: firstly, that too much state education is socially disruptive; and, secondly, that it is economically unnecessary. The first of these criticisms, so common in nineteenth-century Europe and in colonial settings, is now seldom heard in Third World countries (at least not in public). There is widespread acceptance of the principle of education being a basic human right, even if it may prove impossible successfully to implement universal education in practice. However, there is a preference amongst some to maintain an education system that facilitates or at least sustains social differentiation rather than promote equality of opportunity.

The second criticism, that formal education is economically unnecessary, is a matter of very considerable current controversy. This has been an issue in almost all countries, but especially in the poorer countries of the Third World which have generally dismal economic prospects and a rapidly mounting debt burden. The considerable growth of state involvement in economic and social management over the last few decades has created 'distortions' and imposed financial burdens for which financial returns have proved smaller than expected. In particular the massive expansions of enrolments and public expenditure in education have, despite optimistic cost/benefit calculations made at the time of the investments, seemed to yield low public returns.

However, private returns to education at all levels remain substantial, and the political demand for education remains high. In these circumstances, it is argued that the public interest will lie in restricting public expenditure on education to that which is essential for the workings of the economy and cannot be supported by private expenditure. Since private returns to expenditure have been shown to be much higher than public returns, then withdrawal of the state will reduce the public costs without necessarily reducing public benefits, for private expenditure may increase to maintain the overall size and shape of the system. Withdrawal of the state will encourage employers as well as consumers to take a more active role, and further spread the cost. Such a view is generally consistent with an agenda of privatization of public services and greater reliance on market forces for resource allocation, part of the wider ideology of the 'New Right' to roll back the powers and expenditure of the state.

In the Third World this is now a major issue, as the majority of countries, deep in debt and economic crisis, are being encouraged – if not blackmailed – into structural adjustment programmes that involve extensive privatization and 'cost recovery programmes': pushing the costs of education from the public to the private purse, by raising fees or reducing subsidies, rather than outright privatization (World Bank 1988). Even without external pressure, the general trend has been for increasing private contributions, even in countries where tuition fees have been formally abolished but where headteachers can levy uniform fees, sports fees, outings fees, book fees, building fees, and many others. In a book commissioned by the World Bank to review its experience, Gabriel Roth (1987) sought to justify the shift from public to private expenditure as being not only efficient, in terms of improving the quality of education, but also equitable, in that public systems provide a subsidy to the rich, who are disproportionately more likely to be in school at all levels. The poor tend to be in private and low quality schools at present. Most Third World governments find themselves at odds with the ideological premisses of reducing the public involvement in education, but the external pressures to do so and short-term expendients in cost savings have been sufficient to ensure that the voices of dissent on the Right have been given much greater attention since 1985 than they had been given in the previous 40 years.

Despite these voices of dissent from within and beyond the Third World, continuing public and private commitment to formal education remains strong. Expansion of enrolments has slowed in recent years, and in some countries at some levels may even have fallen. This is due mostly to sharply declining economic circumstances, whether short term, as a consequence of drought or major fall in cash-crop incomes, or longer term, as a result of a cumulative downward spiral in Third World economies. It does not normally imply acceptance of any of the theoretical criticisms voiced from the Left or the Right; nor does it imply a general rejection of education by the mass of the population. It is certainly true that policies in individual countries have been affected in some way by the broad international climate of turning to the market and away from the state to manage social expenditure, and adjustments have been made to the education system, but nowhere is there a mass withdrawal from school by private demand or a mass rejection of public expenditure on schools. The changes that there have been have been generally more a matter of emphasis or necessity than of principle. Education remains prominent in the public mind as a positive contributor to development at all levels, from the nation as a whole and to each community and the individual child in school. Despite some falling off in enrolments in a few countries in recent years, public and private demand for education through the medium of formal schooling remains high.

Chapter 2

Global patterns of education

The demand for education has been enormously variable in space and time. The rapid expansion of enrolments in countries with currently high levels of education provision and attainment in Western Europe and North America has really been very recent. The main period of their expansion was in the second half of the nineteenth century and in these countries a level of almost universal enrolment for at least primary school level had been achieved only by the 1920s. At about that time, however, levels of enrolment in countries of what is now known as the Third World were low. The main period of their enrolment expansion has occurred in the second half of the twentieth century, and particularly in the three decades after the Second World War. This was the period of major political change, with the achievement of independence in many countries, and a period of unprecedented global prosperity when an increasing volume of national and international resources was channelled into education by national and international agencies. Despite these expansions, Third World countries taken as a whole still lie far behind developed countries in the quantity and quality of education provided for their populations. Although there are notable differences between major regions of the Third World, it is clear that these are normally less than differences between the developed world and the Third World. This chapter adopts a comparative perspective to describe and explain differences in levels of enrolment and achievement at the global scale. It does so in the light of the issues raised in Chapter 1 that place education and schooling in the context of broader issues in the theory and practice of development.

An internationally comparative perspective immediately raises the problem of comparing like with like. No two education systems are the same, and we need to be clear about which comparisons are required and meaningful. 'Education', as we have seen, is a fairly abstract construct, usually, but not necessarily, incorporating an element of formal schooling, and is not directly comparable from one country or culture to another. 'Schooling', on the other hand, is more readily identifiable and measurable as the number of years a person has spent in school in full-time study, or as the stage of the formal cycle completed. Most schooling systems have a first or primary cycle, a second or secondary cycle and a third or tertiary or higher cycle, and broad comparisons can be made on that basis.

However, even comparisons of schooling as a measure of *input* can prove difficult, both for number of years in school or number of cycles completed. The idea of a school year as a standard unit for comparison is rather elusive, for the length in terms of hours of instruction can vary substantially, and by a factor of more than two, according to data on the length of the school year in a wide range of countries (Table 2.1). Problems of measurement and interpretation are all too evident from such data: in what sense can 800 hours in an Austrian primary school, with a sophisticated learning environment in home and school and highly qualified teachers, be compared with 1290 hours in impoverished Burkina Faso or 444 hours in Bangladesh? 'Number of years in school' introduces similar measurement difficulties: 8 years in one system may involve the same number of hours or school days as 6 years in another. To have 'completed the primary cycle' can mean a very variable number of years in school: in some countries (e.g. Malawi) there is an 8-year first cycle, but in others the primary cycle may be as short as 4 years, though increasingly a 6-year first cycle has been seen as a favoured norm in most regions of the world. A secondary cycle, too, can be of varying length, and in many countries is in practice split into junior and senior secondary cycles each with its own objectives and expectations and very often in different buildings and with different sets of teachers. Increasingly the phrase *basic* education is used to include a primary and a junior secondary cycle of about 9 years altogether (6 + 3) to provide a complete period of schooling for those aged 6–15 years, a normal expectation of enrolment. Each system has its own formal structure against which standard measures of the amount of education provided are estimated, sufficient for broad international comparisons, but with important caveats over detail.

Much more difficult to assess are measures of *output*, both in absolute terms and as comparative measures. The most common measure of output, literacy, is defined internationally as 'those who can, with understanding, both read and write a short simple statement on every day life' (UNDP 1991:195). This is an extremely vague definition, and is subject to much variation in interpretation in national population censuses and in other data collection. Thus international comparisons of literacy rates are particularly sensitive to local definitions of literacy, and need to be treated with considerable caution. They are not necessarily related,

Table 2.1 Length of the school year in hours/year: selected countries

Country	1st Year	3rd Year	6th Year	Country	1st Year	3rd Year	6th Year
Argentina	950	950	950	Guinea-Bissau	1 000	1 000	1 000
Austria	800	800	*	Guyana	1 200	1 200	1 200
Bangladesh	444	715	*	Indonesia A	1 128	1 380	1 503
Benin	1 080	1 080	1 080	Indonesia B	810	1 350	1 350
Brazil	800	800	800	Indonesia C	1 107	1 476	1 558
Burkina Faso	1 290	1 290	1 290	Jamaica	950	950	950
Cameroun 1	1 024	1 024	1 024	Jordan	1 074	1 074	1 074
Cameroun 2	720	720	720	Korea (Rep. of)	816	952	1 088
Cen. Afr. Rep.	885	885	885	Malawi	930	930	930
Chile	1 140	1 140	1 140	Malta	1 120	1 120	1 120
China	1 230	1 320	1 320	Mauritania	1 110	1 110	1 110
Colombia	1 140	1 140	1 140	Nigeria A	1 128	1 128	1 128
Congo 1	900	900	900	Nigeria B	666	666	666
Congo 2	750	750	750	Nigeria C	878	878	975
Costa Rica	800	800	800	Panama	925	925	925
Côte d'Ivoire	1 140	1 140	1 140	Peru	874	950	1 140
Cyprus	980	980	1 260	Philippines	749	1 000	1 200
Ecuador	833	833	833	Senegal	1 176	1 176	1 176
El Salvador	1 000	1 000	1 000	Sri Lanka	740	900	1 200
Equatorial Guinea	713	713	713	Suriname	653	725	725
Ethiopia	1 230	1 230	1 230	Tanzania 1	1 110	1 480	1 480
Ghana	800	800	800	Tanzania 2	907	907	907
Guinea	780	780	780	Venezuela	875	875	875

* Country where the duration of primary education is less than six years. (For those countries where it is longer, only the first six years have been taken into consideration.)

Note: The countries numbered 1 and 2 are those where systems of standard and double vacations exist.
The countries marked A to C are those where a diversity of situations is officially allowed and a relatively large degree of autonomy is given to the States (Nigeria) or Provinces (Indonesia) in the organization of primary education.

- Nigeria: the data supplied are those for the States of Borno (A), Kaduna (B), and Kwara (C).
- Indonesia: the data have been supplied by the Directorate for Primary and Secondary Education of the Ministry responsible (A), South Sulawesi Province (B), and West Java Province (C).

Source: IIEP 1991:11.

even for any one country, to the length of schooling. One recent conclusion is that '[u]ntil we have better ways to assess literacy and numeracy in widely differing contexts, the answers we do have remain only conjectures and political rhetoric' (Wagner 1990:134).

Measures based on formal achievement in examinations or standard tests are much more satisfactory in themselves and as a basis for comparison. Since examination systems tend to be national and specific to particular curricula or schooling structures, comparison is again difficult, even at degree level. Requirements of a B.A. or B.Sc. degree in India, Bangladesh or Pakistan are slightly different from each other, but very different (considerably lower) than these for a B.A. or B.Sc. in, for example, Kenya, Zimbabwe or Botswana, countries which have also experienced the British tradition in higher education. Only where there are standard tests applied internationally can objectively valid comparative measures of output and achievement be used. These measures are difficult to collect and not at all common, but some are available in a limited sense and discussed later in this chapter.

Bearing these measurement issues in mind, this chapter will make international comparisons that will explore differences between the Third World and the developed world, and also within major regions of the Third World. The material will demonstrate that differences in inputs to and outputs from education systems within the Third World, large and important as they are, are less in absolute terms and in their overall significance than differences between the developed world and the Third World, to the extent that both quantitative and qualitative indices in education are major differentiating criteria that set the Third World apart from the First World.

Closing the global gap in enrolments

In the middle of the nineteenth century education and schooling were strongly associated with the then industrializing and urbanizing countries of Europe and North America. Some countries, as suggested by high rates of literacy of bridegrooms and even brides as recorded in marriage registers (e.g. in Sweden), had had very high levels of literacy for many years; others (e.g. France), and these were more common, experienced major increases in literacy in the second half of the nineteenth century (Fig. 2.1). But the overwhelming majority of countries (including Japan) had adult literacy rates of less than 20 per cent in 1850. Between 1870 and

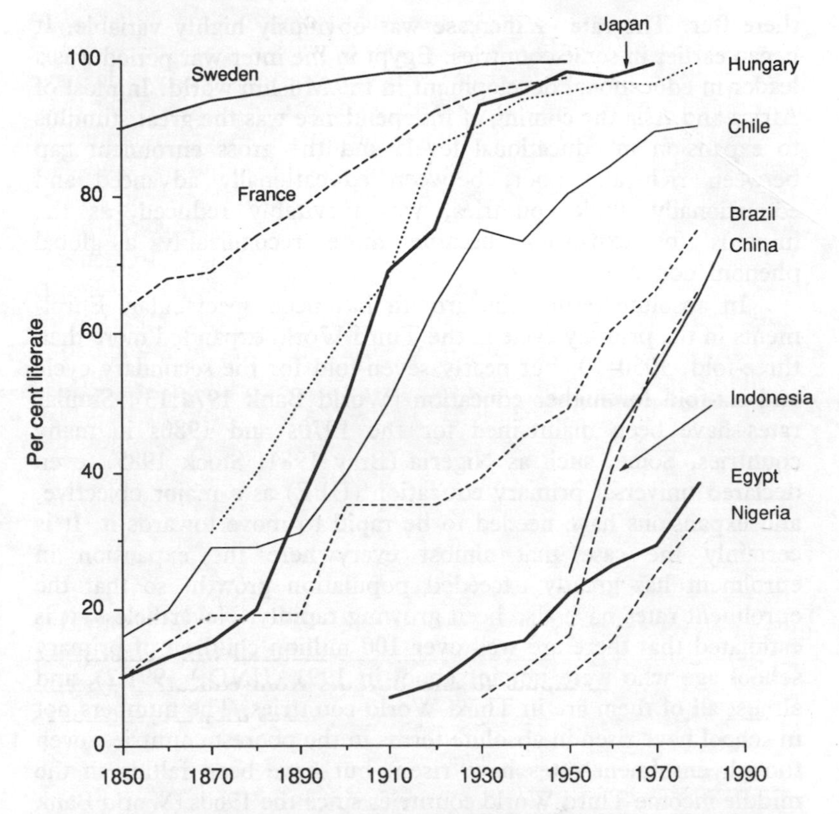

Fig. 2.1 Adult literacy: selected countries: 1850–1985 (from World Bank 1991a:56)

1950 there were major increases in Europe, North America and Japan, to literacy rates normally in excess of 90 per cent by the 1930s.

The countries of the Third World lagged behind in their expansion. The achievement of independence by most Latin American countries in the earlier part of the nineteenth century brought expansions in schooling as indicated by the rising literacy levels in the nineteenth century in Chile and Brazil. Chile experienced a fairly consistent increase between 1850 and 1930, when it faltered, and Brazil saw its most rapid expansion in the 1890s, just after the Republic was declared, though it did not reach the level of Chile. This Latin American pattern up to the 1940s seems to lie somewhere between that of the European countries and that of those countries still or newly dominated by colonial powers. In the latter group of countries literacy levels remained low until after the Second World War, but there were very sharp increases thereafter. The rate of increase was obviously highly variable. It began earlier in some countries. Egypt in the inter-war period was a leader in educational development in the Muslim world. In most of Africa and Asia the coming of independence was the great stimulus to expansion in educational levels and the gross enrolment gap between rich and poor, between educationally advanced and educationally weak countries, was inevitably reduced, as the impetus for expansion became more recognizably a global phenomenon.

In absolute terms, the growth has been spectacular. Enrolments in the primary cycle in the Third World expanded more than three-fold, 1950–70, but nearly seven-fold for the secondary cycle and six-fold for higher education (World Bank 1974:13). Similar rates have been maintained for the 1970s and 1980s in many countries. Some, such as Nigeria (Bray 1981; Stock 1985), even declared universal primary education (UPE) as a major objective, and expansions have needed to be rapid to move towards it. It is certainly the case that almost everywhere the expansion in enrolment has greatly exceeded population growth, so that the enrolment rates have also been growing rapidly. Nevertheless, it is estimated that there are well over 100 million children of primary school age who were not in school in 1991 (UNDP 1991:2), and almost all of them are in Third World countries. The numbers not in school have risen in absolute terms in the poorest countries, even though enrolment rates have risen, but have been falling in the middle income Third World countries since the 1960s (World Bank 1980:107). Despite these major expansions there is still clearly a long way to go before the enrolment gap between the First World and the Third World is approaching closure.

The global pattern of change seems to be strongly related to income level and economic performance of the major regions of the Third World. Table 2.2 identifies gross enrolment ratios at primary, secondary and tertiary levels for standard World Bank income group categories and for regional aggregates. The data give the enrolments of each level as a proportion in the appropriate notional age groups in each country, and are over 100 per cent for some primary level ratios due to enrolment of children outside these notional age ranges. Also in the table are per capita GNP data for these same groupings, with weighted means and the range of values where appropriate. Figure 2.2 describes the proportions in secondary school for 1988 against the per capita GNP for 1989 for those of the 124 individual countries of over 1 million population identified in the World Bank tables for which both education and income data are available.

Table 2.2 Income and educational level by income and regional groupings of countries

	GNP per capita ($) 1989		Percentage of age group enrolled in education 1988				
			Primary		Secondary		Tertiary
	weighted mean	range	weighted mean	range	weighted mean	range	weighted mean
Income groupings							
Low Income Countries (excluding India and China)	300	80–500	75	23–100+	37	4–71	3
India	340	—	99	—	41	—	n.a.
China	350	—	134	—	44	—	2
Lower Middle Income	1 360	610–2 320	103	59–100+	54	13–74	17
Upper Middle Income	3 150	2 450–5 390	104	94–100+	58	38–95	16
High Income	18 330	6 020–29 000	103	71–100+	93	44–100	40
Regional groupings							
Sub-Saharan Africa	340	80–2 960	67	23–100+	18	4–33	2
East Asia	540	180–4 400	128	87–100+	46	27–87	5
South Asia	320	180–430	90	26–100+	37	5–71	n.a.
Europe, M. East and N. Africa	2 180	640–5 350	98	67–100+	60	36–95	14
Latin America Caribbean	1 950	630–3 230	107	83–100+	48	19–82	17
OECD members	19 090	12 070–23 810	105	95–100+	95	74–100+	41

Source: World Bank 1991.

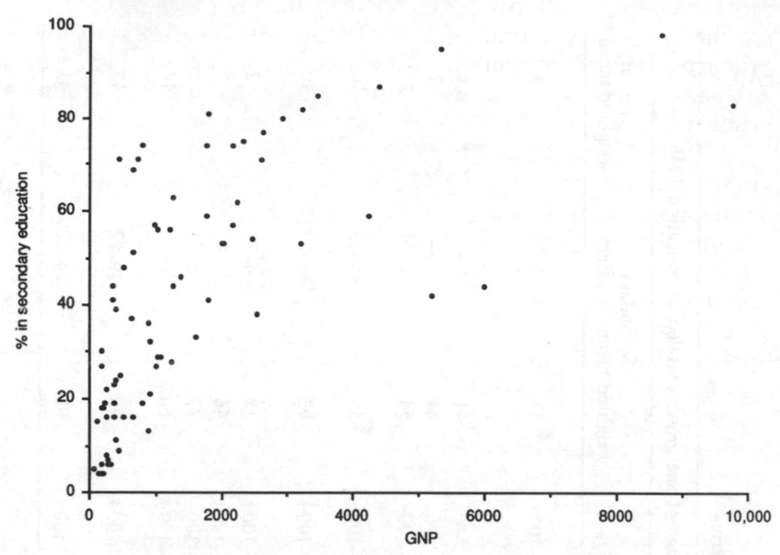

Fig. 2.2 Percentage enrolment in secondary schools, 1988, and per capita GNP, 1989: countries with per capita GNP less than $10 000 (from World Bank 1991a)

The general global relationship between educational development and income level is clear from these data. Richer countries have higher levels of enrolment at all levels than poorer countries, whether in a comparison of all 124 countries or aggregated into income groups. Levels of gross primary school enrolment ratios are near or above 100 per cent in a large number of countries, but mostly in the upper middle and high income group, though even in some low income countries the ratio is over 100 per cent: e.g. Togo, Sri Lanka, and most notably China, where there is clearly a major problem of interpretation of a ratio of 134. Eleven of the 38 countries in the low income category whose primary enrolment ratios are recorded have ratios in excess of 90 per cent; only six of the 40 recorded lower middle income group have ratios of less than 90 per cent. All the upper middle income and high income countries have ratios of over 90 per cent. Clearly there is a ceiling (flexible in absolute terms due to the age-data problem), but the majority of countries have moved towards a universalization of the primary cycle. Only in the poorest countries is there still some way to go at this level. This clearly affects Sub-Saharan Africa most of all, for only here is the weighted mean, at 67 per cent, less than 90 per cent. South Asia, dominated by India and its high ratio (99 per cent), is next lowest with a mean of only 90 per cent, brought down by low proportions for Bangladesh (59 per cent), Pakistan (40 per cent) and Bhutan (26 per cent). Some of the high income countries are the rich oil exporting countries of the Middle East with lower primary school enrolments than their incomes might suggest. Saudi Arabia, in particular, has only 71 per cent of the total age group in primary school and only 65 per cent of the girls.

The range of values at the secondary level between the major groups and regions of the Third World, and between them and the developed world, is more striking. There is a clear positive relationship between the proportion of the age group in secondary school in the low income countries (37 per cent, but with 41 per cent and 44 per cent in India and China, respectively), and 54 per cent in the lower middle income countries and 58 per cent in the upper middle income countries. There is an overall mean of 42 per cent for these groups, compared with 93 per cent for high income countries. Here again Africa lags very far behind the other areas, with the South Asian proportions being raised by a relatively high figure for India, but pulled down by proportions of 18 per cent in Bangladesh and 19 per cent in Pakistan, both of which are large with populations of over 100 million people (Fig. 2.3).

Though the weighted means for the two middle income groups are similar, the range of values is rather different: from 13 per cent (Papua New Guinea) to 74 per cent (Argentina) for the 33 lower middle income countries with recorded data, and from 38 per cent (Brazil) to 95 per cent (Greece) for the 14 upper middle income countries. Three of the lower middle income countries (Senegal, Cote d'Ivoire and Papua New Guinea) have values of less than 20 per cent, and seven of them (Philippines, Dominican Republic, Chile, Poland, Argentina, Bulgaria and Mongolia) are at over 70 per cent. Three of the upper middle income countries (Brazil, Iraq and Oman) have proportions less than 50 per cent, but half the countries (7 out of 14) have values of over 70 per cent. The differences between these two income-based groupings are also evident in the wider range of values in both East Asia and Latin America. In East Asia, Korea has a figure of 87 per cent for secondary enrolment, and is the only NIC to be included in the data (Singapore and Hong Kong are in the high income category, and Taiwan is not separately identified for political reasons) and is in fact the only East Asian country in the upper middle income category. Three others (Philippines 76 per cent, Thailand 28 per cent and Malaysia 57 per cent) are in the lower middle income group. For Latin America and the Caribbean, Haiti (19 per cent) has a similar level to some other Central American republics (Honduras 32 per cent, Guatemala 21 per cent, El Salvador 29 per cent), but some nearby countries, including Jamaica (63 per cent), Dominican Republic (74 per cent), Mexico (53 per cent) and Panama (59 per cent), are at much higher levels.

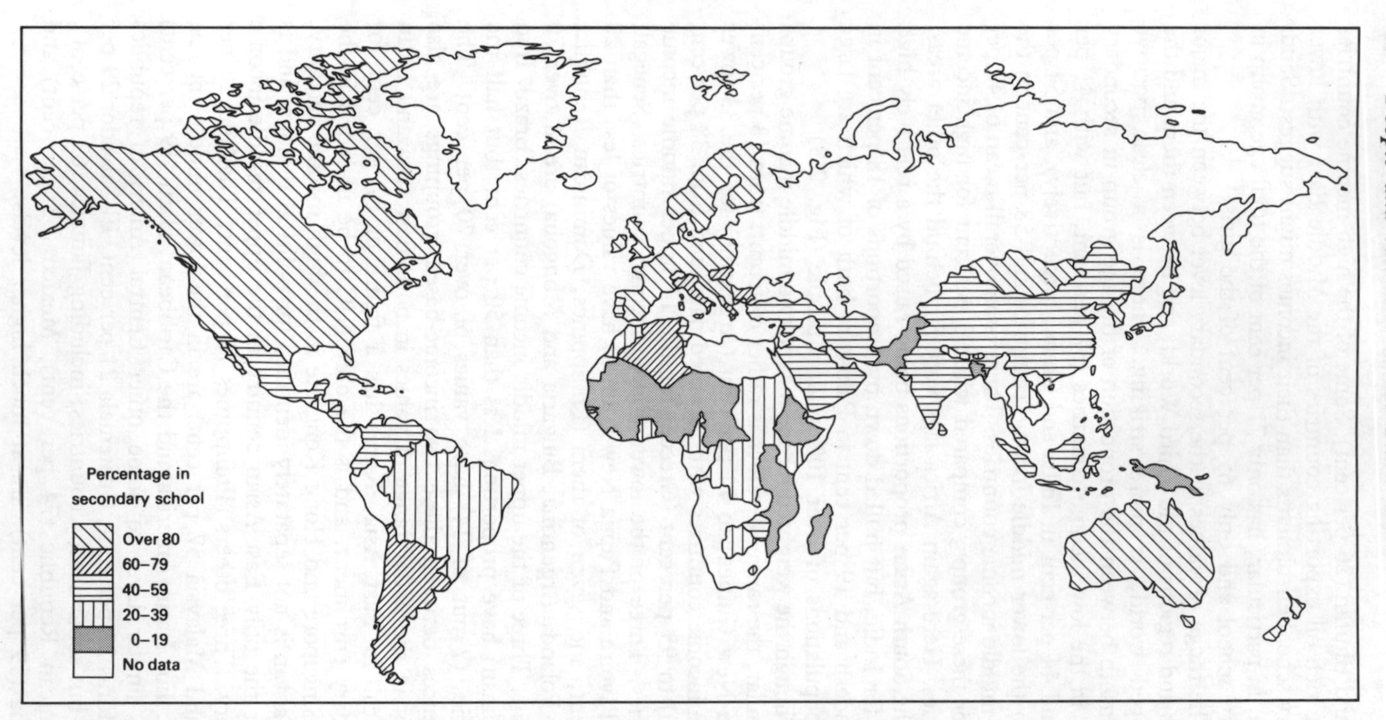

Fig. 2.3 Percentage enrolment in secondary school 1988 (from World Bank 1991a)

At the secondary level, however, the greatest contrast is with the high income countries. In this group Saudi Arabia (44 per cent) and United Arab Emirates (62 per cent) stand out as major anomalies for largely historical and cultural reasons. In most countries, including all the members of OECD, the most industrialized states, proportions are in excess of 80 per cent. These countries can afford high levels of secondary provision, even if levels of demand mean that there is somewhat of a shortfall as some children leave school before completing the secondary cycle. It should be noted that Singapore and Hong Kong, included in this group, have values of 69 per cent and 74 per cent respectively, well below the mean for the high income group.

The disparity between Third World and developed countries is most marked for the tertiary level. While 40 per cent of the age group is enrolled in education in the high income countries, with 41 per cent in OECD countries, the weighted mean for the low and middle income countries is only 9 per cent, with regional values ranging from 2 per cent for Sub-Saharan Africa to 17 per cent for Latin America and the Caribbean. Here there is a substantial difference between enrolment levels in the low income group (3 per cent) and the middle income group (17 per cent), with a very small difference between the means for the upper and lower middle income groups. Even though data for India are not available, it is likely that the level of tertiary enrolment is similar to that of Bangladesh or Pakistan (both 5 per cent), such that the figure for South Asia is probably less than 10 per cent, but above that for East Asia and Africa. These clearly confirm the strong differentials based on income within the Third World. The poor countries cannot afford the massive per capita expenditure that is required at this level. The lack of tertiary education means a shortage of skilled manpower, in itself inhibiting a growth in income levels.

Expansions have also been differential between levels of the educational hierarchy. Using the same income and regional groupings as in Table 2.2, it is possible to explore recent rates of change at each level (Table 2.3). The proportions enrolled at all levels in all groups have risen, often dramatically, between 1965 and 1988. Even in the primary school sector, proportions of the age group enrolled have much more than kept pace with population growth. (If they had done only that, the proportion would have remained constant.) Those areas with lowest starting proportions at that level have had the greatest incentive to expand most, and they have done so. The low income countries saw enrolments rise 53 per cent in this period, from 49 per cent of the age group to 75 per cent, with highest proportional increases in Sub-Saharan Africa. The lower middle income countries had a larger expansion than

Table 2.3 Enrolment by level and income and regional groupings, 1965 and 1988

	Primary			Secondary			Tertiary		
	1965 %	1988 %	% inc	1965 %	1985 %	% inc	1965 %	1989 %	% inc
Income grouping									
Low Income (excluding India and China)	49	75	53	9	25	178	1	3	300
India	74	95	28	27	41	52	5	n.a.	n.a.
China	89	134	51	24	44	83	0	2	n.a.
Lower Middle Income	89	103	16	25	54	116	7	17	143
Upper Middle Income	98	104	6	28	58	107	6	16	167
High Income	104	103	−1	61	93	52	21	40	90
Regional grouping									
Sub-Saharan Africa	41	67	63	4	18	350	0	2	n.a.
East Asia	88	128	45	23	46	100	1	5	500
South Asia	68	90	32	24	37	54	4	n.a.	n.a.
Europe, Middle East, N. Africa	85	98	16	32	60	88	8	14	25
Latin America	98	107	9	19	48	152	4	17	325
OECD	104	103	−1	63	95	51	21	41	95

Source: World Bank 1991a.

the upper middle income countries, which even by 1965 had achieved high enrolment proportions at the primary level. The lower middle income countries by 1988 had caught up with the upper middle income countries. The large reported increase for East Asia to well over 100 per cent of the age group is largely due to the apparently anomalous value of 134 for China (noted above).

In all income and regional groupings, the rate of increase at primary level has been much less than the rate of increase at secondary level. In Africa there was a massive expansion of secondary enrolments from 4 per cent of the age group in 1965 to 18 per cent of age group in 1988, and the proportion in East Asia doubled. Overall the proportion in the low income countries rose from less than 10 per cent to 25 per cent of the age group. However, there were at this level large expansions in the high income countries, with an increase of over 30 percentage points to over 90 per cent enrolment, so that while the percentage increase seemed to have an inverse relationship with level of income and a low initial point, the absolute increase in percentage rate was highest in the richer countries as they moved towards a universal enrolment at the secondary level. Even in Latin America, where the increase at primary level was inevitably small from an already fairly high base in 1965, the secondary rate more than doubled, but to only about 50 per cent of the age group.

At the tertiary level too the proportional increases are very large for the poorest countries, since they start from a very small base to achieve only 3 per cent of the age group enrolled, but in the rich countries the proportions doubled from a mean of about 20 per cent to about 40 per cent. The extent of increase of secondary and tertiary enrolments is also described in Fig. 2.4. Each line represents the change from 1965 to 1988 in secondary proportions (horizontal axis) against tertiary proportions (vertical axis). The length of the line identifies the absolute level of increase. As indicated by the slope of their lines, Africa and East Asia experienced rather greater increase in secondary than in tertiary rates; the Latin American and, even more obviously, the OECD high income countries have proportionately larger increases at the tertiary levels, as there is still further room for expansion at that level, and also the resources and political commitment for that expansion.

The overall pattern described in this section suggests a general progression that gives an early development priority to the primary level, but not until the proportion in school approaches 100 per cent does the emphasis shift to strong growth at the secondary level. Here too there seems to be a tendency for a general progression to secondary enrolments reaching a threshold of over 70 per cent before rapid expansion occurs at the tertiary level.

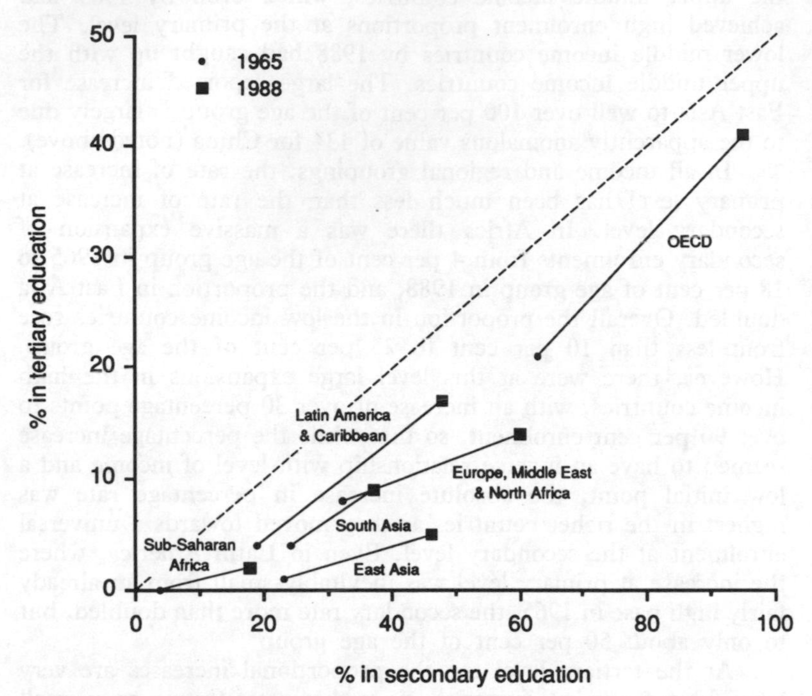

Fig. 2.4 Percentages in secondary and tertiary education, 1965 and 1988, by major regions (from World Bank 1991a)

Girls' enrolment

The increases in overall enrolment have generally meant that the proportion of girls in the schools has risen in most countries. Except in a very few special cases (e.g. Botswana, and Lesotho at the primary school level) more boys than girls have been enrolled. This is a function of both income and cultural preference. The poorer the country, the lower the proportion of girls in the enrolments at all levels, but the boy:girl ratio varies substantially within all income groups. These proportions and changes in them are described in Table 2.4 for the same income and regional groups as previous discussion. In both 1965 and 1988 there is a strong relationship between income level and the proportion of girls enrolled at primary and secondary levels. As the overall enrolments increased between 1965 and 1988 so the relative position of girls seemed to improve in all cases, but not spectacularly.

Figure 2.5 plots the change between 1965 and 1968 in the enrolment rates in secondary school for total population and girls for various groups of countries. In Latin America by 1988 there was a higher proportion of girls than of boys in secondary school,

Table 2.4 Enrolments by sex, income level and regional groupings, 1965 and 1988

	% of age group enrolled in education							
	Primary				Secondary			
	Total		Female		Total		Female	
	1965	1988	1965	1988	1965	1988	1965	1988
Income groupings								
Low income (excluding India and China)	49	75	37	68	9	25	5	20
India	74	99	75	83	27	41	13	29
China	89	134	–	126	24	44	–	37
Lower middle income	89	103	81	101	25	54	22	54
Upper middle income	98	104	94	103	28	58	24	58
High income	93	103	93	100	27	54	25	55
Regional groupings								
Sub-Saharan Africa	41	67	31	60	4	18	2	14
East Asia	88	128	–	123	23	46	–	41
South Asia	68	90	52	76	24	37	12	26
Europe, Middle East, N. Africa	85	98	73	92	32	60	27	55
Latin America	98	107	96	108	19	48	19	52
OECD	104	103	106	103	63	95	61	96

Source: World Bank 1991a.

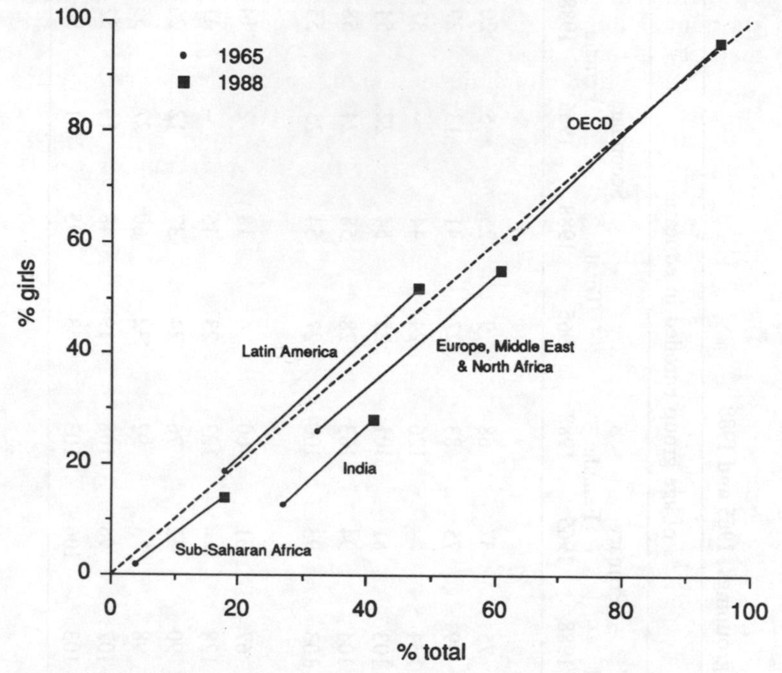

Fig. 2.5 Percentage girls enrolled and total enrolled in secondary school, 1965 and 1988, by major regions (from World Bank 1991a)

that proportion being the same in 1965. Only this case rises above the diagonal of equality in the diagram. For Sub-Saharan Africa, by contrast, the trend line seems to have moved marginally away from the diagonal of equality. Although the proportion of girls in secondary school rose from 2 to 14 per cent, the proportion of boys rose from 6 to 22 per cent, the gap widening from 4 percentage points to 8 percentage points. For India and Europe/Middle East/North Africa, the trend lines run parallel to the diagonal. Figure 2.5 further confirms the general relationship of enrolments of both sexes with income.

Rather different data may be used to examine these income and cultural factors in enrolments of girls. The UNDP *Human Development Report, 1991* contains educational indices in its discussion of female–male differences in the quality of life. It calculates that for all developing countries females as a percentage of males in primary schools rose from 61 per cent in 1960 to 91 per cent in 1987/88, with rises from 43 to 78 per cent in the 40 least developed countries. By 1987/88 the proportion of females in secondary and tertiary enrolments in all developing countries was 76 per cent and 54 per cent respectively. These represent substantial improvements in a period when incomes rose in the majority of countries.

In 1960 the cultural dimension in the male:female ratio seemed stronger than it did in the late 1980s. In particular in the Muslim countries of the Middle East, the low ratios of girls improved dramatically. In Syria girls were only 44 per cent of primary enrolments in 1960, well below the developing countries' average, but this was 94 per cent by 1987/88, i.e. well above the average. Equivalent figures are 26 to 90 per cent for Libya, 38 to 88 per cent for Iraq. Though some rich Islamic countries by 1987/89 still had low proportions of girls enrolled at this level (Saudi Arabia 75 per cent, Morocco 68 per cent), most of the countries with girls at less than 75 per cent of boys are poor countries, the majority in Sub-Saharan Africa (Fig. 2.6). While it is still clear that poor countries that have predominantly Muslim populations (e.g. Pakistan 55 per cent, Guinea 48 per cent, Mali 61 per cent) fare particularly badly, the greater strength of the income factor in girls' enrolments is clear. As countries grow richer they can afford to have more places for girls in school. Then the demand for girls' education may also rise as opportunities for employment in non-domestic and non-subsistence activities also rise, and some education may be required for these. The opportunity costs of girls' education falls.

The increase in enrolments and consequent narrowing of the accessibility gap between boys and girls at all levels in the education system has not been matched by equivalent increases in women's access to jobs. The formal labour force in most Third World countries is still very much dominated by men. The impact of women has been small in absolute terms, though much more in some countries than in others. It is one thing to provide school places; it is quite another to ensure that the theoretically similar education for boys and girls has similar cognitive and occupational outcomes, for these outcomes are highly constrained by broader gender relationships and social attitudes. Even in the schools there may be a 'hidden' curriculum – and sometimes not so hidden – that may effectively reinforce the gender stereotypes of that broader society. As Gail Kelly has argued, using examples from her own work in countries as different as the United States and Vietnam, 'achieving equality in society takes more than opening schools to women' (Kelly 1990:140).

International comparisons of quality

More can mean worse. Even where the absolute level of resources allocated to education is rising there may be a per capita decline if

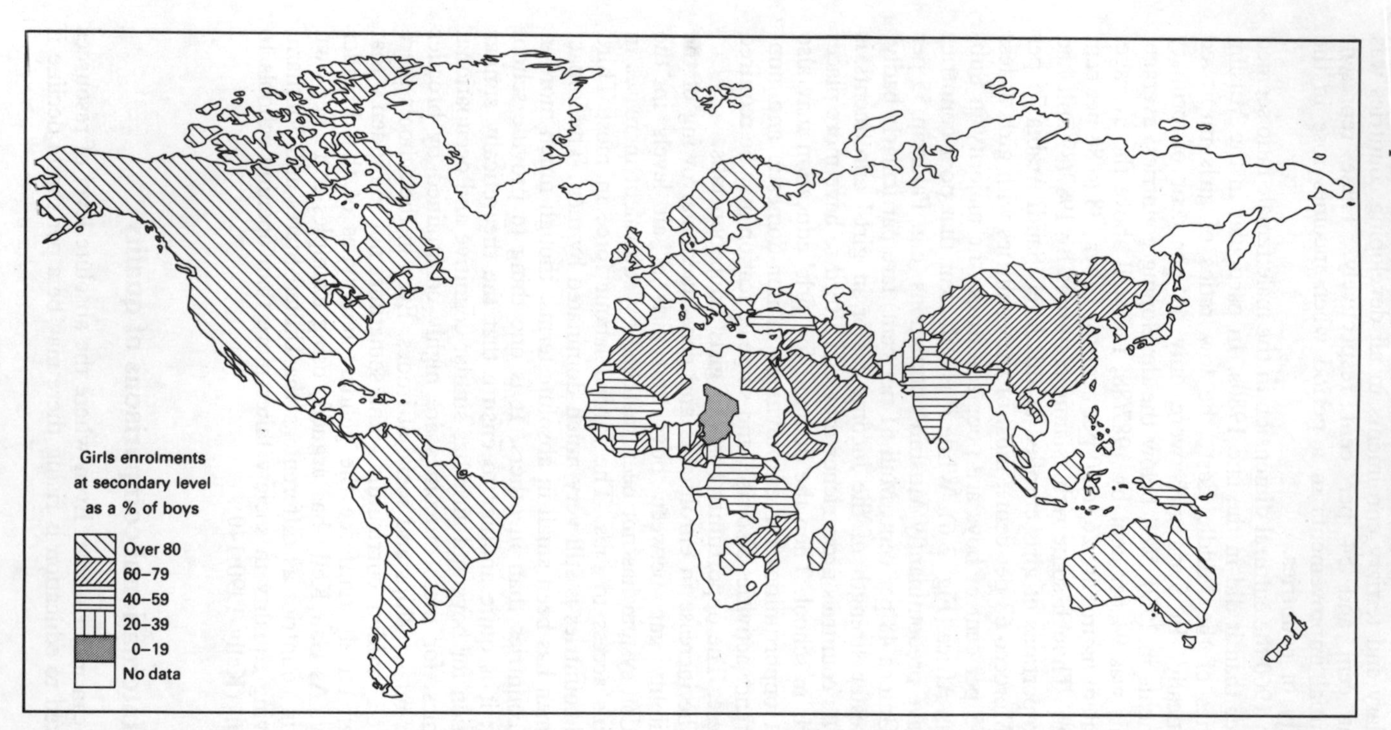

Fig. 2.6 Girls' enrolment in secondary schools as a percentage of boys' enrolment, 1988 (from UNDP 1991)

enrolment expansions are rapid. Equity and the political imperative can easily justify an emphasis on quantitative expansion in Third World countries. However, if per capita resources for education are to fall where there is expansion then quality of education, as measured by inputs of teaching materials, teacher quality, buildings and furniture, and associated outputs, as measured by the amount of learning, may be expected also to have fallen. The quantitative gap between rich and poor countries has been narrowing. However, the amount of resource going into the education system has been rising in the richer countries, even though the proportion of a rising national income going to education has not risen. The increment has been disproportionately devoted to qualitative inputs such as better and more continuous teacher training and other support, to textbooks and a wide range of learning aids and to better buildings and furniture, as well as to lowering pupil:teacher ratios and increasing per pupil space provision in schools. The qualitative gap has been growing.

International comparisons of quality, of both inputs and outputs, are extremely difficult. This is because education systems differ substantially not only in their structure and in the content of their learning, but also in their objectives. The cultural component of education, its social objectives, is least amenable to comparison, indeed to any form of quantitative measure. A system with relatively few and weak qualitative inputs or low levels of attainment or output may satisfy the deeper cultural objectives set for it, that are related to a political function such as citizenship or nation building, rather than to formal learning. This can be a particular problem in comparing education in socialist societies with systems in other countries with less commitment to social engineering through education. Another particular component of quality that is difficult to measure in any context, let alone in international comparison, is the effectiveness of its teachers. This is a critical input that directly affects the achievement of pupils, socially as well as in measured ability in literacy or numeracy.

In terms of obviously measurable inputs, schools in Third World countries are much less well resourced than schools in rich countries. Within the Third World there is a wide and probably growing gap in the level of input (Heyneman and White 1986). Poorest countries spend least, absolutely and proportionately, on non-salary expenditure – mostly books and other learning materials. In the early 1980s teachers' salaries were 96.4 per cent of all recurrent costs in primary schools in Africa, leaving only 3.6 per cent for other costs such as classroom materials. This latter figure was 8.8 per cent in Asia, which includes many richer countries, but 14.4 per cent in industrialized countries. Poor countries spend a smaller proportion of a much smaller budget on quality inputs. The value of classroom materials per student, according to a World Bank study in the early 1980s, ranges from less than $5 in the middle income countries to $75 in Italy and over $300 in the United States. In poor countries pupils must share books, may have no desks or lack even a classroom with a blackboard. Pupils in the United States have more than 140 times the amount of reading material available than pupils in schools in the Philippines, by no means the worst-off country in this respect (Heyneman 1983, 1990). Such gross differences are paralleled to some extent by the qualifications of the teachers themselves. In the Third World there are much lower levels of attainment amongst teachers, both in terms of length of their schooling and length (and quality) of pre-service training, even when they have had any. In many poor countries a significant proportion of the teaching force has been exposed to only a minimal pre-service or in-service training, and many have no formal training at all.

It is hardly surprising therefore that these low levels and poor quality of inputs are reflected in gross differences in the level of achievement. International comparative measures of achievement have been pioneered in particular by the International Project for Evaluation of Educational Achievement, the so-called IEA studies. These were at first mostly in developed countries, but have now extended into many Third World countries (Keeves 1988; Purves 1987). The general methodology and premisses of the IEA studies is a subject of much controversy amongst educationalists (see Postlethwaite 1987, for a positive, insider's view, and Theisen *et al.* 1983, for a trenchant criticism), but for the present purposes they are heavily dependent on skills tests, such as in mathematics and science, that have a very strong Western bias. The researchers involved in the IEA structure are, rightly, very wary of an 'Olympic Games' approach to their results that provides only national rankings, but one of the most general findings is the positive relationship between mean national achievement scores in a range of tests (e.g. on reading comprehension, science, mathematics) and national per capita income. What is also clear, however, is that within the sample, inter-school dispersion of scores about the mean scores is higher in poorer countries than in richer. This is probably related, at the inter-school level, to gross differences in resource allocations (see below, Chapter 4).

Expenditure on education

These sharp contrasts in quality and quantity may be directly related to ability to pay. Most education in Third World countries is provided by the state as a direct charge on government expenditure, and inevitably poor countries will spend less in absolute terms than rich countries. In proportional terms, by comparison, there is no clear link between GNP and educational expenditure. UNDP (1991) estimates that educational expenditure amounts to 3.7 per cent of GNP for the developing countries, taken as a whole. The average for Sub-Saharan Africa is 3.8 per cent of GNP, with a range of values from 8.5 per cent (Zimbabwe) to 1.4 per cent (Nigeria) and 1.7 per cent (Tanzania). Figures are rather lower overall for the low income countries of South Asia – Bangladesh (1.3 per cent), Pakistan (2.2 per cent) and India (3.4 per cent). On the whole they are higher in the middle income countries, with exceptional proportions of over 10 per cent in 1986 in Saudi Arabia and Libya, but more typical proportions in Costa Rica (4.5 per cent), Chile (5.0 per cent), Malaysia (7.9 per cent), and South Korea (3.0 per cent). In other words there is a fairly wide range of values that in large part may be attributable to measurement problems, not least of which is the proportion of GNP that is spent by government.

When government expenditure is the yardstick, then it seems that expenditure on education has been falling, even though the proportion of GNP has been rising. Expenditure by Third World countries was an estimated 2.2 per cent of GNP in 1960 compared with 3.7 per cent in 1986. In World Bank data on the share of government expenditure devoted to education, on the other hand, the proportion has generally fallen. Sometimes the fall has been very large, as in Bolivia where it fell from 31 per cent of government expenditure in 1972 to 11 per cent in 1986. In Tunisia it fell from 30 to 14 per cent in the same period. In low income countries overall the fall was from 15 to 10 per cent, and in middle income countries from 20 to 14 per cent in that period. The discrepancy is due to the rapidly rising share of government expenditure that is now needed to service debt repayments, low income countries in the 1990s now being net exporters of capital (Heyneman 1990). There is a very large squeeze on govermnent expenditure, and education has suffered, though perhaps less than some other sectors such as health services. The proportion of GNP devoted to health expenditure in developing countries rose from 1.0 per cent in 1960 to 1.4 per cent in 1986, much less rapidly than the proportion on education.

Significant though these levels of expenditure in the Third World are, they have to be seen in the context of absolute expenditure on a global scale. The countries of the North not only spend much more on education, for they have more people enrolled, but they spend much more per capita. Here again comparative data are hard to find and often difficult to interpret, but UNDP has been able to calculate a mean per capita education expenditure for the North, and to compare it with per capita expenditure in 37 Third World countries (Table 2.5). The mean for these 37 countries is 27 per cent of the North's expenditure. In some of the richer Latin American countries per capita expenditure is near or even exceeds the mean of the North. In eight countries per capita expenditure is less than 20 per cent of that in the North. Africa fares least favourably in such a comparison, with a mean of 18 per cent, and the three South Asian countries in the table all have figures in the 20s. Such a disparity in resources for education between the North and the Third World begins to explain differences in levels of achievement and 'quality'. In many senses, the wonder is that so much is achieved with so little!

Conclusions such as these provide strong further support for increasing the flow of resources to education in the Third World, and particularly to the poorest countries. They suggest the importance of the improvement of quality as a means to enhance equality as well as to raise economic performance. In the rich countries there is a fairly weak link between school and parental occupation, for home socio-economic status is generally a better predictor of occupational status than level of schooling. In poorer countries, however, the role of the school seems to be more prominent. Schooling, even in the poorest countries with low levels of quality inputs and low levels of achievement in international terms, does provide children with experience and skills that differentiate some children from others, more than the background effect of their home circumstances does. That experience will greatly affect their life chances in the modern economy (Heyneman 1980b, 1989).

Global issues

The evidence considered in the previous sections of this chapter has been firmly set in a global context to consider broad continental patterns of enrolment, quality and expenditure. At one level it suggests an optimistic picture of very substantial improvements in initial and continuing access to schooling in Third World countries. More resources have been made available, more children – both boys and girls – have been able to stay for longer in school. At another level, however, it suggests a less optimistic interpretation of a great deal still to be done, with still high proportions of children out of school and, for those that are in school, relatively few and declining resources available to them. As a result overall levels of learning and achievement have been disappointing. Most

Table 2.5 Index of real expenditure per capita on education, 1985: North = 100 (selected countries)

	<20	20–39	40–59	60–79	80+
Sub-Saharan Africa	Zambia Nigeria Malawi Ethiopia Mali	Zimbabwe Kenya Madagascar Cameroun Côte d'Ivoire Tanzania Senegal	Botswana		
South Asia		Sri Lanka Pakistan India			
East Asia	Indonesia		Hong Kong Korea Philippines		
Latin America and Caribbean	Honduras Bolivia	Brazil El Salvador	Argentina Columbia Paraguay Peru Dominican Rep.	Ecuador Guatamala	Uruguay Chile Venezuela Panama
Middle East, North Africa		Morocco	Tunisia		

Source: UNDP 1991.

of all it is clear that the gaps in enrolments, resources and achievement between the Third World and the developed world are very wide, and seem to reflect the broader structures of that global inequality that gives the Third World its basis as a valid generalization and sets it apart from the rich countries of North America, Western Europe, Japan and Australasia. Furthermore, when the effects of the 'brain drain' are taken into account, the net flow of highly qualified and educated personnel from the Third World to the First World (Chapter 7), the inequalities in the global distributions of educated people are further accentuated. Najafizadeh and Mennerick (1988) have gone as far as to describe the world-wide expansion of education in the period 1950–80 as a 'failure'.

Education in this perspective is seen as a general feature of the processes of development that sustain the continuing vast inequalities in levels of income and quality of life between the 'North' and the 'South'. It is an integral component of the processes of underdevelopment as well as being apparently necessary for development. As such, education cannot be entirely separated from broader global issues, such as trading relationships, resource flows or industrial development, that are at the heart of the global development debate. It is an international concern, but views about education and its role in development range very widely. The views from and of the Third World countries might be expected, given the range of data presented in early sections, to be very different from those from and of the rich countries. In global perspective, education has a broad range of meanings, uses and values.

That education is a global issue is well evidenced by its role in the United Nations, and explicitly with the 'E' in UNESCO. At the time of its establishment immediately after the Second World War as a leading specialist agency of the newly founded United Nations, the United Nations Educational, Social and Cultural Organization was instrumental in ensuring that education remained at the forefront of the global agenda. It developed and promoted the standardization of data collection procedures and analysis for education systems, and it developed a role in facilitating global comparison. It had a role in policy analysis and formulation, and was able to exercise great influence on Third World countries at a time in the 1950s and 1960s when many of them were becoming independent of colonial control and seeking to restructure their existing education systems to meet the demands of what seemed to be a new economic and political order. As was noted in Chapter 1, this was pursued in the UNESCO continental education conferences of the early 1960s that committed most countries to universal primary education and set targets for achieving that goal, and in the establishment of specialist institutions in educational research and planning. UNESCO played a major role in setting the global educational agenda for a period of perhaps 25 years, 1950–75. That agenda was dominated by the positive view of education as a basic human right, as defined in the UN Declaration of Human Rights. Furthermore, since UNESCO came to be dominated by Third World interests, it sought to narrow the global gap, to pursue policies of expansion of enrolments in the first place, but also to ensure quality improvements through such measures as curriculum change and textbook development. This was essentially a 'liberal' and 'modernizing' view of education that saw the educational experience of the West, apparently successful for economic and social improvement, as needing to be extended in large measure to Third World countries. The objectives of expansion – more of the same, quality improvement, more resources, better teachers – were largely taken for granted.

By the 1980s, however, the global agenda for education was being set largely by the World Bank and the IMF. This meant an ideological shift to an approach dominated by economics and economists rather than by politics, sociology and educationalists (Altbach 1988). The Brandt Commission (1980) had relatively little to say about education directly, except that it should be more 'relevant', i.e. to the needs of Third World countries, but also that it should benefit from additional resource allocations from the global redistributions of wealth that were at the heart of Brandt's recommendations. These additional resources would be targeted at the rural and urban poor, and the recommendations were consistent with the basic needs strategy being pursued by the World Bank in the 1970s. The value of this equity thrust could hardly be denied, nationally as well as internationally, and levels of inequality at all scales seemed to be falling with massive enrolment expansion. However, most countries found it difficult to generate sufficient resources internally to finance the inputs (both capital costs and recurrent costs) that were needed, for economic performance in many countries had faltered.

Resources therefore needed to be obtained from external sources. UNESCO has no direct resources of its own to support efforts of individual countries. Despite UN recommendations on fixing levels of bilateral aid as a fixed proportion of GNP of rich countries, the amount of bilateral aid in general, and even more so for education, was falling. By contrast, the amount of resource available from multilateral sources was rising. Since education was not seen by commercial banks to be a productive investment in their terms (i.e. with a fairly short period for return on investment), countries tended to seek assistance from multilateral institutions such as the Asian Development Bank, the African Development Bank and most of all from the World Bank. The World Bank, as has been discussed, has been well disposed to investments in education since the 1960s. The justifications for that support within the World Bank, even in its basic needs phase, were largely in terms of efficiency rather than equity. In an organization dominated by economists, and operating in association with its sister Bretton Woods institution, the International Monetary Fund, it was inevitable that efficiency in terms of value for money that is familiar to an economist should come to the fore. All the better if there could be equity objectives too, but these were no longer priority criteria by the late 1980s.

Countries anxious to expand education and improve its quality sought the support of World Bank finance, but were as a consequence required to expose their policies and practices to World Bank scrutiny and conditionality. The World Bank has its own policy priorities and preferences and its resource strength means that it is able to exercise very great influence on the direction of educational change of Third World countries, through specific project loans and also more generally in broad programme support. In the education sector the World Bank is able to offer and ensure the implementation of advice that is entirely consistent with its broader strategies of structural adjustment programmes that gained momentum during the 1980s (Mosley *et al.* 1991).

The approach of the World Bank to structural adjustment in education, to make the education system more efficient and provide better value for scarce resources, is epitomized in its 1988 Policy Study, *Education for Sub-Saharan Africa: policies for adjustment, revitalization, and expansion.* Using its considerable experience of lending for educational development in almost all Sub-Saharan African countries, the staff members of and consultants to the World Bank developed a diagnosis of considerable pessimism. Expansion had been rapid in the period immediately after independence (i.e. in the 1960s), and this was much needed. However, there was insufficient change in the curriculum and in the links between education and the changing needs of national economies: too much expansion with inadequate resource support so that quality fell. Furthermore, by the 1980s African economies were extremely weak, rocked by internal shocks (drought, famine, civil war) and external shocks (oil price rises, declining terms of trade), so that internal resource allocations were falling, absolutely and per enrolled child, at a time when population was growing rapidly. Enrolment ratios were by the mid-1980s stagnating, in some countries even falling. Public educational expenditure was generally falling as a proportion of all government expenditure. Having reached an average of nearly 5.5 per cent of national income in 39 African countries in 1980 (compared to 6 per cent in OECD countries), expenditure had fallen to nearly 4 per cent by 1983. However, that level was still higher than the average for all

Third World countries, which had been consistently rising between 1970 and 1983, but was still below 4 per cent of national income on average.

The study recommended that the fall in resource allocations should be halted, indeed reversed. At the same time there need to be significant reductions in per capita costs at all levels. At primary school this may require raising pupil:teacher ratios; at secondary school the increased use of day rather than boarding schools; at tertiary level better management of universities and other institutions. Generally, however, public unit costs need to be reduced by more 'cost sharing': i.e. passing more of the direct costs of education to the consumer through higher fees and other user costs. Given the presumption that expenditure in education is 'rational' and productive and that the benefits accrue to the individual being educated as well as to society as a whole, then the beneficiaries should be able to pay and be prepared to pay.

Furthermore, since quality needs to rise substantially at all levels without incurring commensurate increases in expenditure, then more innovative delivery systems need to be developed. Much weight is given in the World Bank study to distance learning techniques (radio programmes, correspondence courses, etc.) and to better and cheaper availability of learning materials through market pricing. Obviously the study sought to explore how quality could be raised at the same time as costs were being reduced ('efficiency'), and these are not likely to be easy bedfellows. Policy recommendations that formulate specific measures to achieve these objectives have been implemented, at least partially, in some countries (e.g. Ghana, Zambia, Senegal) as part of broader economic adjustment measures that have given rise to such critical comment in Africa. The financial clout of World Bank support has ensured that reforms that would normally be difficult for governments to implement, or even discuss publicly on purely political grounds, have been put in place.

The main problem with this approach is that it is not consistent with equity criteria. By pushing a higher proportion of costs on to the user, the poor within each country tend to be even more marginalized within the system, for they can least afford the extra expenses. It is certainly true that in the past the better-off have benefited disproportionately from public support for the costs of education, for it is they who disproportionately reach secondary and tertiary levels, the most subsidized sectors. To reduce the extent of that subsidy will mean that the well-off may continue in school, even if they have to pay more. The poor may have no alternative but to withdraw altogether. The net effect of these structural reforms in education is likely to be socially regressive, at least in the short term, while economic activity remains at a low level. It is recognized even within the World Bank that structural adjustment programmes will not promote equity in the short term, but they will promote efficiency, and this is a necessary prerequisite for medium- and longer-term redistributive policies targeted at the poor. The adverse impact on the poor is recognized in such programmes as PAMSCAD (Programme to Mitigate the Social Costs of Structural Adjustment) in Ghana, in which funds are made available to vulnerable groups for community activities. Grants in the education sector have mostly taken the form of funds to communities for the material costs of construction of classroom and laboratory blocks in junior secondary schools.

The growing importance of the World Bank builds on the role pioneered by UNESCO in bringing greater uniformity and international comparability between education systems. There is less scope for individual governments to develop their own particular ideas. Unless they are able to finance them from internal sources, they will be subject to externally imposed conditions that require the use of fairly traditional models derived largely from the Western experience. Innovation and experimentation are not easy, even in the most favoured circumstances, and certainly not encouraged in the financial environment of structural adjustment. Standard criteria for preparing and appraising proposals and for evaluating outcomes push the trend further towards uniformity, to put in place reforms that can be similar in many Third World countries (Rondinelli *et al.* 1990). In theory this should make cross-national comparison easier; it should move national experiences towards a similar pattern. However, it would be no surprise to find that national experience of the impact of education, both economically and socially, continued to differ, even within the constraints imposed by World Bank criteria. A systematic comparative education perspective on national and regional differences within the broader, externally imposed agenda will continue to provide important insights.

The global issues identified in this chapter find an echo in each country, but their elaboration and relative importance vary considerably. Any global perspective on contemporary education and development relationships is thus modified by the ability of national states to impose their own strategies and outcomes on these overarching forces. For many poor states, however, that power is weak and the external dominance is a source of great resentment. The need for local capacity building in education is strongly apparent in three particular respects: in human development; in the restructuring of public and private institutions to allow them to operate effectively; and in ensuring a political leadership that allows these institutions to be carefully nurtured (King 1991). Local capacities could permit the development of not only a Latin American or African or Asian or Arabic model, but a Salvadorian, Somalian, Syrian or Sri Lankan model. It is to the national structures, capacity, assumptions and objectives that attention will now turn.

The geography of educational provision: the national scale

Education systems are essentially national. The national level is a natural scale at which to consider the nature and functioning of the system and how it affects the life and work of the population within the area of the state. With very few exceptions (notably Switzerland and, in some respects, the United States) the essential structures of the education system, its length and curriculum, the qualifications and salaries of teachers, and the means of finance are all set nationally. Though there can be important differences within countries, and many of these will be explored in this chapter, they are constrained by the national framework set for the system by the sovereign governments. As was argued in the previous chapter there are large and important differences between countries, even neighbouring countries, in the structures and organization of education, such that inter-country differences are likely to be much greater than intra-country differences. Yet it was also argued that, as a result of external pressure from multinational lending and planning agencies, notably the World Bank and UNESCO, these international differences in both quantity and quality have been reduced in recent decades. Despite this trend, they remain greater than differences within most Third World countries. The range of experience examined in the previous chapter in terms of crude levels of enrolment or of quality in terms of measured achievement, or of levels of expenditure is normally much less between each region but not necessarily between each school within each country.

Nevertheless, the internal variations that do arise within countries are important to our understanding of the relationships between education and development, and must therefore figure largely in the concerns of this book. In any country the education system fulfils a range of functions, all of which cannot be dealt in this chapter with in anything like adequate detail. However, we begin by a brief consideration of the educational system as an economic, political and social phenomenon, before turning to fuller consideration of the national education system as a geographical phenomenon: one that functions within the national space to serve the national population.

Schools as economic phenomena

Like any other industry, the education system is part of the economic system of any country. As with agriculture or manufacturing, it consumes resources and generates a product: the educated person. It employs people, and sacks them. It uses physical plant, in the form of schools and other institutions. In a strictly functional sense 'schools are to the education industry as factories are to industry at large' (Richmond 1975:5). Education is traditionally part of the service sector, and as such can be assessed as an economic phenomenon. However, because of its importance to the economies of Third World countries, not least because it consumes a large proportion of government expenditure, its direct and immediate economic significance is even greater than it is in the developed countries. This significance is felt principally in the numbers employed, in the construction industry and in the manufacture of supplies for classrooms and the pupils themselves.

The teaching force in most Third World countries is large as a proportion of those employed in the formal sector of the economy. It is often more than 50 per cent of those employed in the public sector, itself a high proportion of total employment. Furthermore, teachers as well-educated people themselves are amongst the more highly paid groups in the community, certainly above mean income levels, often by a large factor. In Sub-Saharan Africa in the mid-1980s the mean salary of a primary school teacher was 5.6 times the mean per capita GNP. For a secondary school teacher it was 10.6 times (World Bank 1988:147). The level and structure of teachers' salaries is normally a contentious issue. Teachers themselves often feel they are very badly paid and as a result are poorly motivated, both in absolute current levels and in terms of historical levels. In their own expenditure patterns and in terms of what is expected of them by others, e.g. in using imported consumer goods and living in 'modern' housing, their salaries are generally low and have been falling in real terms. Yet since they earn salaries well above the average, certainly far above the ratio of teachers' earnings to average earnings in developed countries, the argument is often levelled that they are overpaid. There are strong pressures, not least within government, to keep teachers' salaries in check.

The most obvious impact of relatively high salaries is apparent in rural areas where teachers may be by far the largest contributors to the local economy in expenditure on everyday 'low threshold' goods such as foodstuffs or personal needs – soap or medicines – or in larger expenditures such as house building or land purchase. Since schools tend to be well spread into rural areas, so too are teachers, even if there is usually a difficulty in attracting teachers to work in remote rural schools and persuading them to stay there. Partly as a result of their own higher incomes, but probably also because of their own higher levels of education and experience of the commercial economy, many teachers are themselves employers and business people in the money economy. They augment their earnings as small businessmen or women, and own such facilities as small shops or vehicles such as lorries or small 'pick-ups'.

Schools themselves provide a major stimulus to the construction industry. Building standards for schools are normally higher than for most domestic buildings in rural areas, and are constructed on a commercial basis by local or larger companies, financed from local community or government sources. Even where schools are built using community labour, the construction industry will provide many materials, specifically for roofing and often bricks or roof supports. The commercial component is likely to be higher for secondary schools than for primary schools, for these have more extensive buildings, are built to higher standards and with more sophisticated needs for water, sewerage and other services.

In addition to the buildings, schools need to be supplied with a wide range of services and facilities provided from within the country, perhaps from one centralized and nationalized depot, as in the case of several countries where the Ministry of Education exercises a virtual monopoly over the preparation and/or production and/or distribution of most books and other materials used in the classroom (e.g. Tanzania, Zimbabwe). At the other extreme are countries where textbook production and distribution are entirely within the private sector. In these countries (e.g. Malaysia, India, Kenya) there are many bookstores which depend as retail outlets on the trade generated by the schools as institutions with bulk purchases and by individual pupils. The normal case is to have a mix of both, with the Ministry directly providing base textbooks and other materials (e.g. exercise books), but with a parallel private sector offering alternatives and supplements to that centrally organized, but often incomplete, base. The strength of private demand for education can often ensure the existence of a private school bookshop and supplies outlet even in small settlements. In Kenya many of the government sponsored industrial estates and Rural Industrial Development Centres set up in small towns find that a textbook depot or small-scale printer manufacturing exercise books is a familiar type of 'industry' attracted.

The same may be said of furniture, particularly desks, for in some countries their provision is a central responsibility and provided under contract by large-scale manufacturing concerns. This is the case in Botswana where primary school furniture is distributed by the Ministry of Local Government and Lands, responsible for the running of the primary schools. In other cases furniture is a local responsibility and a staple product of very variable quality of small local concerns, usually using locally available materials. This can be a problem in many countries in savanna or desert areas where wood is in short supply, and the possibilities for a local school furniture industry are so much less than they are in forested areas. In the several countries of the coastal forest zone of West Africa, desks and other furniture can be much more readily found in schools compared with the interior savanna and sahel, or with countries such as Lesotho or Ethiopia or Bolivia and other areas of the altiplano in the Andean countries of South America where there is wood but it is in such great demand as domestic fuel that only alternative, often imported, materials are possible for school desks.

Many schools and parents display a strong attachment to school uniforms. Here again these may be, as in Indonesia, a direct monopoly of the Ministry of Education at least in providing the design; in other cases the responsibility will be with the school to identify the material, but with local tailors to supply it and to make up the tunics, blouses, shirts or trousers. Many tailors in rural areas rely substantially on school uniforms for their trade, and in many rural training and vocational schools the making of uniform items is a popular and financially attractive feature to generate additional income for the school.

Schools in rural areas can make an important contribution to stimulating local tradespeople and entrepreneurs. Their role as economic phenomena that can allow them to make that contribution is all too often restricted by centralized decision making and purchasing that tends to favour the larger scale and urban-based suppliers. However, the expenditure of teachers and of larger institutions like boarding secondary schools or teachers' colleges can have major impact on small towns. Even some larger centres, such as Zaria in Nigeria (Van Raay, 1970) or Cape Coast in Ghana, which are the location of several major educational institutions, are dependent on these institutions and the expenditure of their staff and students for some of the economic vitality of the centres. In these and other similar cases of towns which grew rapidly as administrative centres in the colonial period, the morphology of the town has been much affected by the large areas of land taken up for school campuses. In India and Pakistan many of the long established schools and colleges occupy large sites in the *cantonments* area of towns.

Schools and the political system

Given the economic importance of education in both its short-term local impact as described in the previous section and its longer term potential for public and private advance, it is hardly surprising that the education system should be seen as an intensely political phenomenon. It is the focus for political ambition and rivalry, an arena for the working out of wider political issues in many Third World societies, including (probably especially in) those countries where formal party politics is banned or restricted. The school in these circumstances becomes the place where many of the wider concerns and rivalries in society at all levels are aired and fought out.

More generally, however, education is never far from the public attention. It is a regular headline concern in newspapers – often more concerned with troubles in schools or financial irregularities than with strictly educational issues. It is a subject of continual debate in bars and coffee houses: who has achieved what and where, who are the new teachers, which is the better school? This is further evidence of the strength of that private demand catalogued in Chapter 1. Most households are involved at one level or other, and the effects of schooling can be critical to the future of the household.

This allows schools to be important vehicles with which aspiring politicians like to be involved, either as a teacher or headteacher, or as a community leader or member of the board of management or a school governor. A politician can make his mark locally and at a wider scale by offering a better schools system in the first instance, and delivering these promises once elected or appointed to a position of power. All politicians will promise new schools or better facilities in their constituencies, and anyone who can deliver such a promise is clearly a 'big man' of influence. Politicians everywhere prefer to allocate resources in their own interest, and since a high level of resources of the Third World governments are allocated to education, it becomes a key means of patronage. In Nigeria in the 1960s, in a period of great expansion of secondary education after independence and at a time of mounting inter-regional political tensions, the Federal Government established three new national secondary schools whose locations were: 'intended to foster political integration. The three ministers who decided their location came from Sokoto, Warri and Afikpo: the three schools were allocated to Sokoto, Warri and Afikpo' (O'Connell 1966:137).

Patronage can involve not only the location of new buildings, but also the appointment and allocation of teachers and other resources. All this of course lends to rivalries which may be played out in the schools system, with headteachers taking sides to the benefit or detriment of their schools, creating serious, often career- or even life-threatening rivalries within management bodies. The theoretical ideal of a politically neutral school, catering equally for the needs of all the community, and being given and using resources on much the same basis as other schools and communities, is often far removed from the reality on the ground, even to the extent, as in parts of Pakistan, of schools having known political allegiances and these allegiances affecting parents' choice of school for their children.

The political role of schools is greatest at secondary and higher levels, where not only are the institutions more prestigious and command more resources, but the students are politically aware and politically active. School strikes and other similar events are common in many countries, bolstering the political tradition and prominence of schools generally, and political figures are often involved in both the creation and solution of political tensions within the institution. At university level students may offer the only vociferous opposition to the government and there is a long tradition of student 'troubles' in many countries. The education system at all levels is so important that it cannot be relegated to the status of a technically apolitical and non-controversial service akin to sewerage or sanitation (and even in these apparently less controversial services there are major political issues, e.g. in urban squatter settlements).

Schools and society

At its broadest, schools have a role to play in the socialization of children, to prepare them to be members of a local community and national society. Even in pre-modern societies there were usually institutions, though of a less formal sort, for 'initiation' and providing a mechanism outside the family or household for introducing young people, often boys and girls separately, to their expected contributions to society. Modern formal schools equally provide a mechanism, outside the immediate confines of the family or clan, for introducing children to a wider community, to allow them to learn to work and co-operate with other children. In addition they subject children to an authority and regime within the school that is different from that at home. They are exposed to a general awareness of people, events and circumstances beyond their immediate environment. School and family are complementary elements in the wider socialization of children.

The social effects of schools affect the life and career patterns of those who have been to school, and in particular their social mobility. In earlier years, when schools were relatively few and few people had formal education, schooling was indeed a necessary, perhaps even a sufficient condition for rapid social mobility. The few educated people moved into niches in the newly created social order of colonies: in the bureaucracy, in commerce and in the political sphere. An educated élite seemed to have been created and this was a major factor in stimulating the private thirst for education. However, as schooling became more generalized, its important role in social mobility and social change has lessened. The élite perpetuate their social status at the expense of the poor by ensuring that their children attend the best schools that are well equipped and with well-qualified teachers. It is because the education system in most countries reflects society so well that it often provides the fascination for students of the sociology of education: it holds up an easily identifiable mirror in which wider social processes can be reflected (Foster 1977). Measures which seek to promote social mobility by directly targeting the marginalized and disadvantaged do not always succeed. In Ecuador, for example, externally financed projects of consolidation (nuclearization) of schools to improve quality and of non-formal education to bring more adults into the national economic system failed to gain the support of the national political élite or of the expected beneficiaries, the poor of remote Andean comminities, and the projects failed to meet their objectives. Preston (1985) argues that this was because the projects targeted the symptoms of inequality rather than its fundamental causes, which were to be sought outside the education system.

One area in which the social constraints are particularly apparent is in girls' education. In schools, as in society as a whole in many Third World countries, there is a systematic bias against women that has longstanding cultural roots, and may be further reinforced through modernization. In Ethiopia, for example, it has been shown that amongst the reasons why girls achieve less well than boys in school are family background variables and greater domestic commitments (Abraha *et al.* 1991). In most countries enrolment of boys is considerably in excess of that of girls, and increasingly with higher levels of education. Often there are as many girls as boys starting in the first year of school, but girls drop out more readily, in part because of greater demands on their time for domestic chores, in part because the long-term economic return is less. Girls have fewer opportunities in the modern sector labour market, and in any case as mothers they continue to have major domestic responsibilities. Botswana and Lesotho, where more girls than boys attend the early stages of primary school, seem to be major exceptions. The anomaly is due to the specific labour demands for children in pastoral societies such as Botswana where young boys look after the family herds of cattle, usually at some distance from the family's permanent home in a village, and hence at some distance from a school, and to the specific needs of the South African mining industry for unskilled and uneducated Lesotho labour. By the time of secondary school, however, even in Botswana and Lesotho boys outnumber girls in the classroom.

The reflection of wider processes in society is not merely in terms of levels of enrolment, but also in the ways in which in many – but not all – Muslim countries there are separate boys' and girls' schools, often physically next to each other. This differentiation finds its reflection in the status of women in society as a whole. The broader structure may also be manifest in different curricula for boys and girls – practical skills for employment or self-employment for boys; domestic skills for girls, again confirming, indeed exacerbating the social *status quo*. Wider issues about changing the status of women, for example by bringing them into the labour market, may be first fought out in the schools in controversies over the introduction of practical subjects for girls: which subjects and at what level?

Schools and culture

As noted above, there have been traditional institutional arrangements in most societies for the transmission of the culture of society – its values and religion, its music and other arts, its view of itself and of others – but the transmission of culture has been to a very considerable extent within the realm of family and kin responsibilities. Certainly there have been times when young people are gathered together for ritual or ceremonial purposes such as initiation or for weddings or funerals, and this may be done as a socially selective basis, as in the 'schools' for pages, sons of the nobility, at the court of the Kabaka of Buganda (Watson 1969). In small-scale societies learning about culture was not normally institutionalized, but rather a part of everyday learning in the household and in the community.

With the coming of larger scale societies through colonialism and subsequently with the growth of the modern state, the schools system assumed very considerable importance for the development of a national culture that was very often quite different from the many very different smaller scale and often mutually antagonistic or traditional cultures. In particular the national and colonial culture was often expressed in the national language, generally the colonial language rather than an indigenous language. This national language was the language of instruction in the schools. In Latin America Spanish or Portuguese was strongly associated with the role of the Church as a cultural force. So schools became central to the promotion of national culture, both in the curriculum in history or citizenship classes, and in children's music, drama and other activities. Often there is a Ministry of *Education and Culture* that is now responsible for the schools system. In such countries (e.g. Ghana 1986–89) the cultural role of education has been made explicit.

Yet here lies a threat to traditional cultures, for the national schools system may seek to promote the national culture as part of the nation's attempt to create and advance its national identity at the expense of local cultures. In Africa in particular, governments have been reluctant to sanction the use of other than a very few vernacular languages in the schools and have imposed centralized control over curricula. In Ghana only 15 Ghanaian languages are recognized for use in schools, but over 40 separate languages are used in the country. The schools system supports national unity and a trend towards cultural uniformity rather than cultural diversity. The recent trend in Ghana towards more decentralized structures of planning and administration in government generally, as in the education sector, may leave room for more diversity than was apparent in the years immediately after independence in 1957, but still the education is seen as being more important for the creation of a national culture than for the revival or even survival of traditional cultures.

The education system as a geographical phenomenon

Since it operates over a national territory and provides a service that is targeted to the children in the population in that national territory, the education system is a geographical phenomenon. It creates patterns in space, and is affected by processes that operate over space. Schools are normally the points of delivery of that service, though through radio, television and correspondence courses the delivery mode may be variable. Whatever the form of delivery, the whole aggregates to an integrated and complex system, usually centrally controlled, which is as much part of the geography of any area as the network of roads or farms or the distribution of manufacturing industry.

Any geographical system comprises three elements: nodes, hierarchies and interactions:

1. The *nodes* in this case are the delivery points, the schools. They have their own catchment areas, determined not only by distance – children attending their nearest school – but by a host of other factors which will vary in their importance from place to place and school to school. They include such factors as the quality of education, real or perceived, that is provided, transportation arrangements, if any, for the journey to school, admissions criteria operated by headteachers or other officials, and the economies of scale in educational provision at any given level (Chapter 4).
2. The *hierarchy* is evident in the different levels of the education cycle. There are many widely distributed primary schools, and often they are arranged administratively into clusters with feeder schools for the lowest classes sending pupils to a larger, central upper primary school. There are fewer secondary schools, and in countries with relatively low enrolment ratios at that level they tend to be in towns, at levels of the national urban hierarchy above those lower order centres where there are many primary schools. At higher levels there may be specialized or tertiary institutions, usually in or near the major towns. Where there is only one such institution in the country (e.g. a specialist skill training establishment) it is often located in the capital city. The system tends to be planned and administered hierarchically, to mirror the national administrative structure with central, provincial and/or district and/or local administrations, each with different powers and responsibilities.
3. *Interactions* are the flows of pupils and teachers, as well as information and materials between the nodes and up and down the hierarchy as the system is integrated into a cohesive whole. The major flows are of pupils, from home to school in daily mobility, but also from one school to another at the end of each cycle to a higher level of the hierarchy. Pupils also move out of the system at various points as school leavers to become integrated with the wider national urban hierarchy as a factor in controlling their subsequent mobility in the country, and, as will be explored in Chapter 7, there is a strong relationship between the point at which pupils leave the education system and their subsequent patterns of migration. Teachers also move within the system in transfers, usually to schools at the same level of the

hierarchy but there are often major issues surrounding the posting of new teachers and the transfer of existing teachers in order to achieve a desired distribution. Other flows within the system tend to be top–down, such as the distribution of textbooks and other learning materials and the application and enforcement of centrally agreed policy decisions on curricula, examinations or other administrative function.

These geographical features are summarized in Fig. 3.1, 'the education system as a spatial hierarchy'. The nodes, the schools, serve spatially discrete areas at least in principle, even though in practice many catchment areas even at the lowest level are overlapping. The many primary schools at the lowest level of the hierarchy are well distributed through the population to be accessible to the children on a daily basis. In many Third World countries there are often incomplete schools at the primary level (i.e. they do not offer the whole cycle) so that many pupils will go to full primary schools at a higher level of the hierarchy. In other countries, the minority, there may be formal arrangements at this level with central schools offering the whole cycle surrounded by incomplete feeder schools. In other cases there may be no feeder arrangements, and all primary schools, however small their total enrolment, are expected to cater for pupils at all grades of the cycle. In most countries there are large numbers of drop-outs at the end of each cycle, and often at the end of each grade in the first cycle. Since internal economies of scale for each school increase up the hierarchy, schools become much larger and there are therefore fewer of them. At the higher level there are specialist institutions, in many cases only one per country, and these include universities, teachers' colleges and a range of specialist training and technical institutions, each of which has a national catchment.

The main flows are of pupils upwards within the system and out of it into the labour market, particularly at the end of each cycle, especially where there is a terminal examination for that cycle which determines progression to the next level. As the base of Fig. 3.1 suggests, most flows at the lower levels are to the nearest school, but admission procedures to secondary schools are sometimes based on national or regional criteria that necessitate wide and overlapping catchments. This is even more common for institutions at the tertiary or higher level. The diagram also suggests flows of teachers out of teachers' colleges, normally but not exclusively, at the tertiary level, back into the schools. Since teachers' colleges normally specialize in training of either primary or secondary teachers, the flow is not hierarchical. Trained teachers will go to either secondary or primary schools. Teachers in post will also move within the system through transfer, and there will

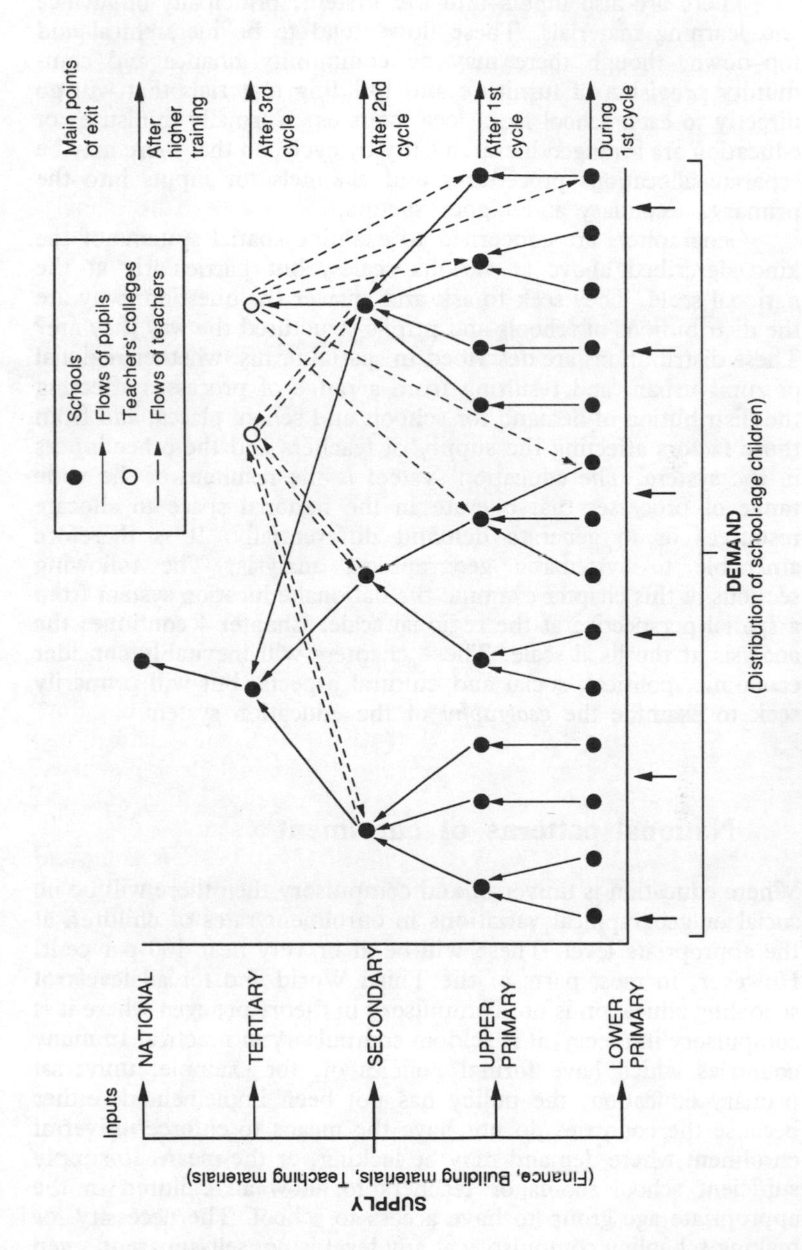

Fig. 3.1 The educational system as a spatial hierarchy

normally be flows between schools at the same level of the hierarchy.

There are also inputs into the system, principally of finance and learning materials. These flows tend to be hierarchical and top–down, though there may be community finance and community provision of furniture and building materials that will go directly to each school from local sources. Normally ministries of education are arranged hierarchically by cycle, so that there may be separate allocations procedures and channels for inputs into the primary, secondary and higher sectors.

Geographers are concerned to examine spatial systems of the kind described above at various scales, but particularly at the national scale. They seek to ask and answer the question: why are the distributions of schools and pupils structured the way they are? These distributions are described in spatial terms, whether regional or rural–urban, and resulting from a range of processes affecting the distribution of demand for schools and school places, and from those factors affecting the supply of teachers and the other inputs in the system. The education system is the resultant of the wide range of processes that operate in the national space to allocate resources or to generate demand differentially. It is therefore amenable to systematic geographical analysis. The following sections of this chapter examine the national education system from a spatial perspective at the regional scale. Chapter 4 continues the analysis at the local scale. These chapters will inevitably consider economic, political, social and cultural aspects, but will primarily seek to examine the *geography* of the education system.

National patterns of enrolment

Where education is universal and compulsory then there will be no social or geographical variations in enrolment rates of children at the appropriate level. These will be at or very near 100 per cent. However, in most parts of the Third World and for all levels of schooling education is not compulsory in theory or, even where it is compulsory in theory, it is seldom compulsory in practice. In many countries which have formal policies of, for example, universal primary education, the policy has not been implemented, either because the countries do not have the means to enforce universal enrolment where demand may be lacking, or the means to supply sufficient school rooms or teachers to allow all children in the appropriate age group to have access to school. The necessity for making schooling compulsory at any level is not self-apparent, even if governments could afford to provide it. Though demand is generally high (as was argued in Chapter 1), there may be good reasons for cultural or economic resistance to modern schooling. Parents may be unwilling to expose their children to the new value system that seems to be inculcated in schools. More likely in poor societies where children from early years make a significant contribution to the household and family economy, the opportunity cost of sending a child to school may be too high, and governments may be reluctant to face the political as well as the economic costs of enforcing attendance. Inequalities in enrolments are therefore the norm at all levels throughout the Third World. The higher the level, the lower the overall national enrolment ratio and the greater the extent of social and geographical inequality is likely to be.

The most common measure of levels of current provision is the enrolment rate, whether measured in *net* enrolment rates (numbers in a given age group as a proportion of the children in that age group), or in *gross* enrolment rates (numbers in a given cycle as a proportion of children *in the notional age* for that cycle). Net enrolment rates are much more valuable indices, but are more difficult to calculate due to the lack of accurate data on the age of pupils in the schools. These comparative rates may be further refined in related indices such as boy:girl ratios or cohort progression ratios derived from disaggregation of enrolment data by sex or grade. They may also be associated with qualitative differences (e.g. examination results or qualifications of teachers). Disaggregated historical patterns of enrolments are often indicated by age-specific census data which specify educational status (e.g. last grade achieved, level of literacy). These may indicate not only substantial changes over time in educational status of the population but also differentiation by region or socio-economic group.

Social inequalities have been a major concern of analysts and governments. The rich tend to receive much more education than the poor, but in a situation where some receive no education at all, the gap is between the rich, who tend to go to school, and the poor, who disproportionately do not. This difference is increasingly marked up the educational ladder with very restricted access by the poor to higher education. Many studies in all parts of the Third World have catalogued how and why social inequalities in enrolment originate, and, even more important in the long term, how and why they have persisted in the face of considerable expansion and considerable efforts to narrow the social and economic gap (Foster 1977). Even in socialist societies, such as China, where the schools system has an overtly social engineering function to contribute to the wider policies of society to narrow and eventually eliminate social class differentials, success has been limited (Lewin and Xu 1989). Since the overall message of studies in the sociology of education is that the schools system reflects wider forces of power and wealth in society, it is hardly surprising that there should be inequalities in enrolments in the schools.

Geographically, patterns of enrolments are also likely to reflect geographical patterns of power and wealth and to be consistent with long established spatial patterns in any country. The roots of contemporary rural–urban inequalities in Latin America are to be sought in the colonial period in the early role of the Catholic Church in provision:

> Schools, seminaries and universities were central features of the towns and cities from which the colonial territories were controlled and administered. . . . Tiers of interlocked and hierarchical responsibility extended from national to regional to local scale of territorial administration, and in each of these levels from urban to rural. As a consequence of élitist educational philosophy, oppressive political control and economic constraint, the ideals of universal education enshrined in the republican constitutions in Latin America still have to be realised in most countries of the region even at primary level. Indeed the incapacity of these systems, after about 150 years of independence, in respect of serving the mass of the population as well as the complex needs of national economies, provides a sobering example to younger developing countries with high expectations of education in terms of development.
>
> (Brock 1985:3)

In Nigeria there have been major differences in provision between the north and south of the country since the early colonial period, differences that have not been eliminated by national policies of universal primary education or allocation mechanisms within federal and state budgets to promote narrowing of inter-state and intra-state inequalities (Falola 1989; Okafor 1989). Higher enrolment rates tend to be found in the richer regions, the highest normally in the administrative region containing the capital city. They tend to be lowest in poorest regions furthest from the political and economic core of the country. The patterns can be described at various scales of disaggregation, usually the administrative units of the country and using either census data (i.e. from questions on those currently in school) or data collected in the schools themselves. Census sources tend to inflate current enrolments as well as the overall educational status of the population, but in many countries the statistical work of the educational ministry is subject to the several sorts of errors (of under-counting in some cases; of double counting in others) such that calculation even of reasonably accurate enrolment rates (i.e. without the problem of reporting of age) can be subject to error.

Regional patterns are traditionally described in choropleth maps, as in the case of primary school enrolments in Ghana (Fig. 3.2). The enrolment data are derived from the annual school census and aggregated to each of the 110 administrative districts and also to ten major administrative regions. The notional primary school

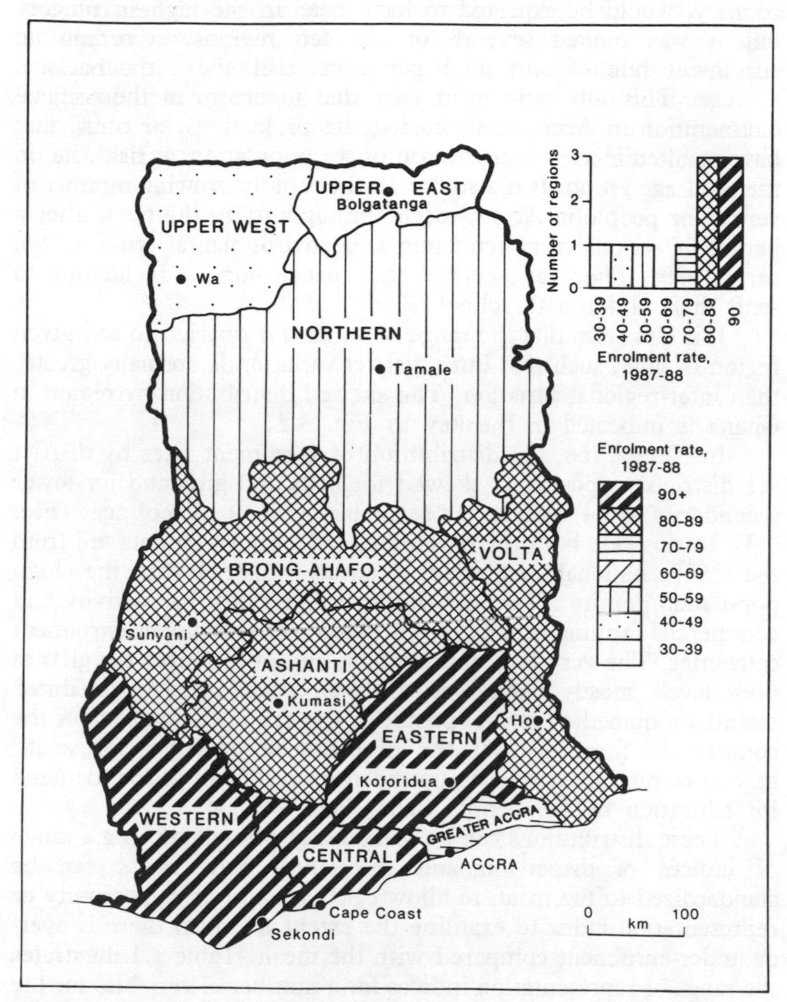

Fig. 3.2 Primary school enrolment rates by region: Ghana 1987/88

age range in Ghana is 6–11 years and the population in that group for 1988 was projected from 1984 population census data. The national mean rate was calculated at 80 per cent, with regional rates of below 50 per cent in three regions (Northern, Upper East and Upper West) and over 90 per cent in two regions (Eastern and Central). In general the south has much higher rates than the north, and the overall pattern of enrolments at primary level, but also at other levels, mirrors the overall pattern of econmic development in the country (Aryeetey-Attoh and Chatterjee 1988). Greater Accra, almost entirely urban and the richest part of the country, would be expected to have been in the highest category but it was ranked seventh of the ten regions, recording an enrolment rate of only 77.5 per cent, still above the national average. This low ratio is in part due to errors in the original enumeration in Accra or to inadequate projections, or both, that have resulted in an overestimation of the population 'at risk', i.e. in the 6–11 age group. It is also due to the rapidly growing number of very poor people in Accra, recent immigrants to the town after a period of serious rural economic and environmental crises in the early 1980s. They have neither the finance nor the inclination to send their children to school.

It is clear also that the range of values for districts in any given region is wide, such that inter-district variation is normally greater than inter-regional variation. The skewed distribution by region in Ghana is indicated in the key to Fig. 3.2.

In Kenya, too, the distribution of enrolment rates by district (41 districts) is positively skewed for boys and girls and for lower secondary (10–14) and upper secondary (15–19) school ages (Fig. 3.3). In this case both the enrolment and population data are from the 1979 national population census. Districts in the high population density areas near the main areas of urban activity and commercial farming occupy the proportions in the upper enrolment categories. The very long tail for boys and even longer for girls at both levels mostly comprises the poor rural districts inhabited mainly by nomadic pastoral people in the dry north and east of the country. In these districts the problems of provision for small, mobile communities are compounded by problems of low demand for education by the people themselves.

These distributions can be statistically compared using a range of indices of dispersion and inequality. Raw ratios can be standardized to the mean to allow calculation of an opportunity or representation index to examine the extent to which there is over- or under-enrolment compared with the mean. Table 3.1 illustrates the range of representation indices for a number of variables for the 16 districts of North West Frontier Province, Pakistan. It is clear from the table that there is a greater range of inter-district values for higher levels of the schools cycle than for lower, and for girls than for boys. The greater range of values is an indication of a greater level of inequality. For all three levels and for boys and girls separately the greatest over-representation is in Peshawar District, capital city of the Province, and for Mardan and Kohat, relatively developed rural districts. By comparison, the most isolated and least developed districts, Swat, Dir, Chitral and, outstandingly, Kohistan, all in the Hindu Kush and Karakoram mountain areas, are under-enrolled at most levels relative to the rest of the province.

The extent of inequality in any distribution can be described

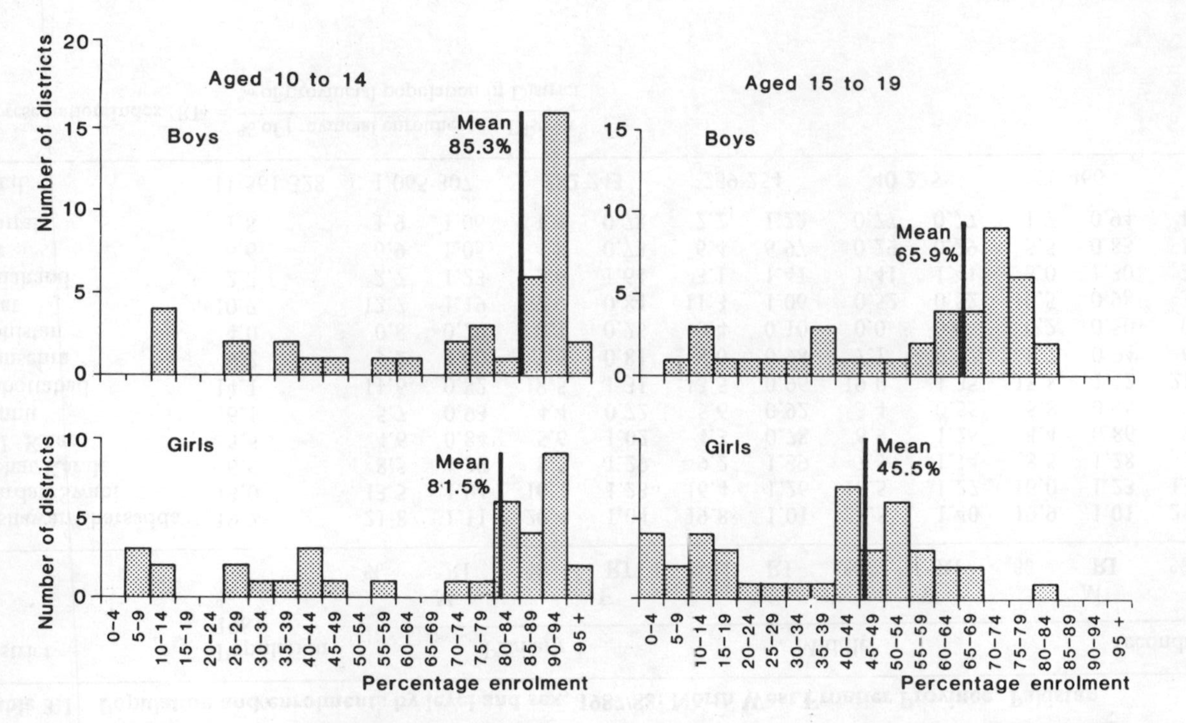

Fig 3.3 Enrolment rates by ages 10–14 and 15–19 and sex by district: Kenya 1979 (from Gould 1988b:44)

Table 3.1 Population and enrolments by level and sex, 1987/88: North West Frontier Province, Pakistan

District	Population 1981 %	Primary				Middle				Secondary			
		M		F		M		F		M		F	
		%	RI	%	RI	%	RI	%	RI	%	RI	%	RI
Peshawar/Charsadda	19.7	21.8	1.11	20.4	1.04	19.8	1.01	27.5	1.40	19.9	1.01	28.7	1.46
Mardan/Swabi	13.0	13.5	1.04	16.1	1.23	16.4	1.26	16.5	1.27	16.0	1.23	13.7	1.05
Kohat/Karak	6.6	8.5	1.29	8.5	1.29	9.2	1.39	7.5	1.14	8.5	1.28	7.3	1.11
D.I. Khan	5.5	4.6	0.84	5.6	1.02	4.3	0.78	6.9	1.26	4.4	0.86	9.4	1.71
Bannu	6.1	5.7	0.93	4.4	0.72	5.6	0.92	3.4	0.55	5.8	0.95	3.8	0.62
Abbottabad	14.1	11.6	0.82	18.5	1.31	13.5	0.96	19.0	1.35	15.8	1.12	21.6	1.53
Mansehra	9.2	9.2	1.00	7.5	0.81	8.0	0.98	7.1	0.77	8.7	0.94	6.9	0.75
Kohistan	4.0	0.8	0.20	0.1	0.25	0.4	0.10	0.0	0.0	0.2	0.50	0.0	0.0
Swat	10.7	12.7	1.19	9.0	0.84	11.3	1.06	0.52	0.52	10.5	0.98	3.6	0.34
Malakand	2.2	2.7	1.23	3.6	1.64	3.1	1.41	1.41	1.41	3.0	1.30	2.7	1.22
Dir	6.6	6.9	1.05	4.8	0.73	6.4	6.97	0.29	0.29	5.5	0.83	1.0	0.15
Chitral	1.8	1.9	1.06	1.4	0.78	2.2	1.22	0.77	0.77	1.7	0.94	1.4	0.78
Total	11 561 328	1 065 307		282 243		239 254		40 225		79 466		11 850	

$$\text{Representation Index (RI)} = \frac{\text{\% of provincial enrolment in District}}{\text{\% of provincial population in District}}$$

in a Lorenz diagram and summarized in a gini coefficient. This index needs to be used with care as it is extremely sensitive to the number of elements in the distribution. Thus it cannot be used to compare the extent of inequality between two countries or within two districts. It can be used, however, as for example in the case of Thailand, to illustrate how the extent of inequality rises up the education system, from a coefficient of 0.033 for lower primary schools (i.e. hardly any inequality with roughly the same proportion of enrolments as children at risk in each of the seven major divisions of the country), to 0.532 for admission to higher education (Fig. 3.4).

The gini coefficient can also be used to examine trends in the

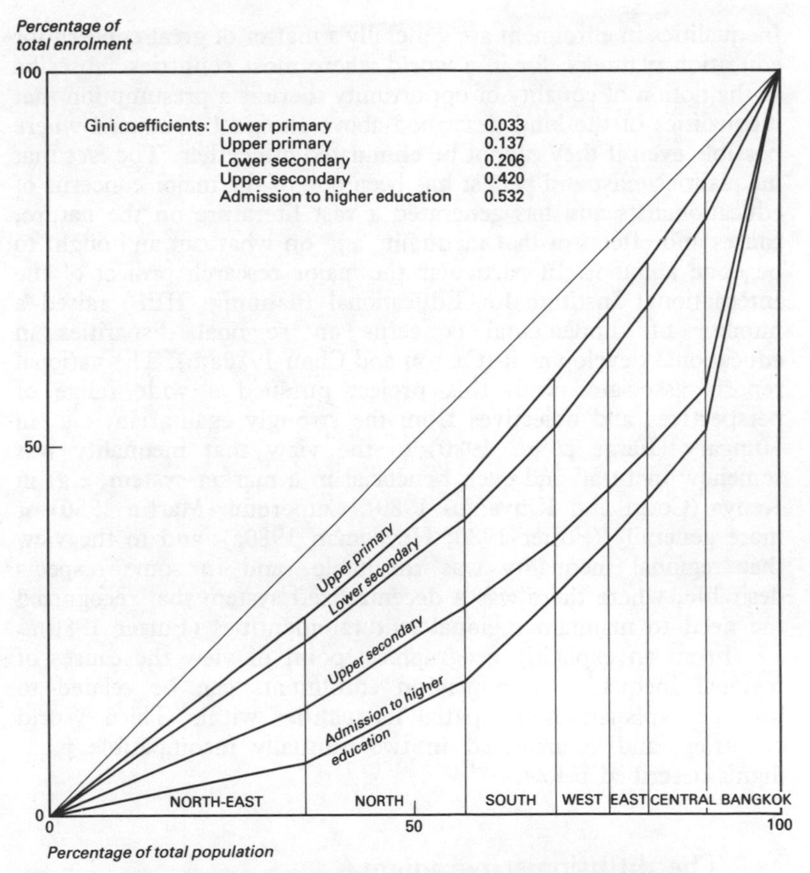

Fig. 3.4 Percentage of population and percentage enrolment by education level and region (Lorenz curves and gini coefficients): Thailand 1976 (from Sudaprasert *et al.* 1980:264)

extent of inequality over time. In Papua New Guinea in the 1970s for example, in a period of rapid expansion of enrolments, the gini coefficient for primary school employments 'improved' from 0.24 in 1971 (then only 18 provinces) to 0.17 in 1974 (by then 20 provinces) and to 0.10 in 1980 (also 20 provinces), indicating a narrowing of inter-provincial inequality (Fig. 3.5). This has been a normal feature in the early stages of national enrolment expansion where there is systematic concern for raising enrolments in previously underserved areas.

Explanations of national patterns of inequality

Inequalities in enrolment are generally a matter of great concern for education planners, for in a world where most countries subscribe to the notion of equality of opportunity there is a presumption that inequalities of the kind described above should be reduced where possible even if they cannot be eliminated altogether. The fact that inequalities exist and persist has been one of the major concerns of educationalists and has generated a vast literature on the nature, causes and effects of that inequality and on what can and ought to be done about it. In particular the major research project of the International Institute for Educational Planning (IIEP) raised a number of fundamental concerns on regional disparities in educational development (Carron and Chau 1980a,b). The national reports associated with this project pursued a wide range of perspectives and objectives from the strongly egalitarian, e.g. in Hungary (Ferge *et al.* 1980) to the view that inequality was somehow 'natural' and even beneficial in a market system, e.g. in Kenya (Court and Kinyanjui 1980), Cameroun (Martin 1980) or more generally (Foster 1980; Heyneman 1980a), and to the view that regional inequality was inevitable, and in some respects desirable, where there was a decentralized system that recognized the need to maintain regional cultural identities (Furter 1980).

From an explicitly geographical point of view the causes of regional inequality in education enrolments can be related to broader explanations of spatial inequalities within Third World countries, and summarized in two mutually incompatible paradigms described below.

The diffusionist paradigm

Neo-classical theories of development have emphasized the modernization of the Third World as a spread of Western technology, economy and ideas as a spatial process from the West into the

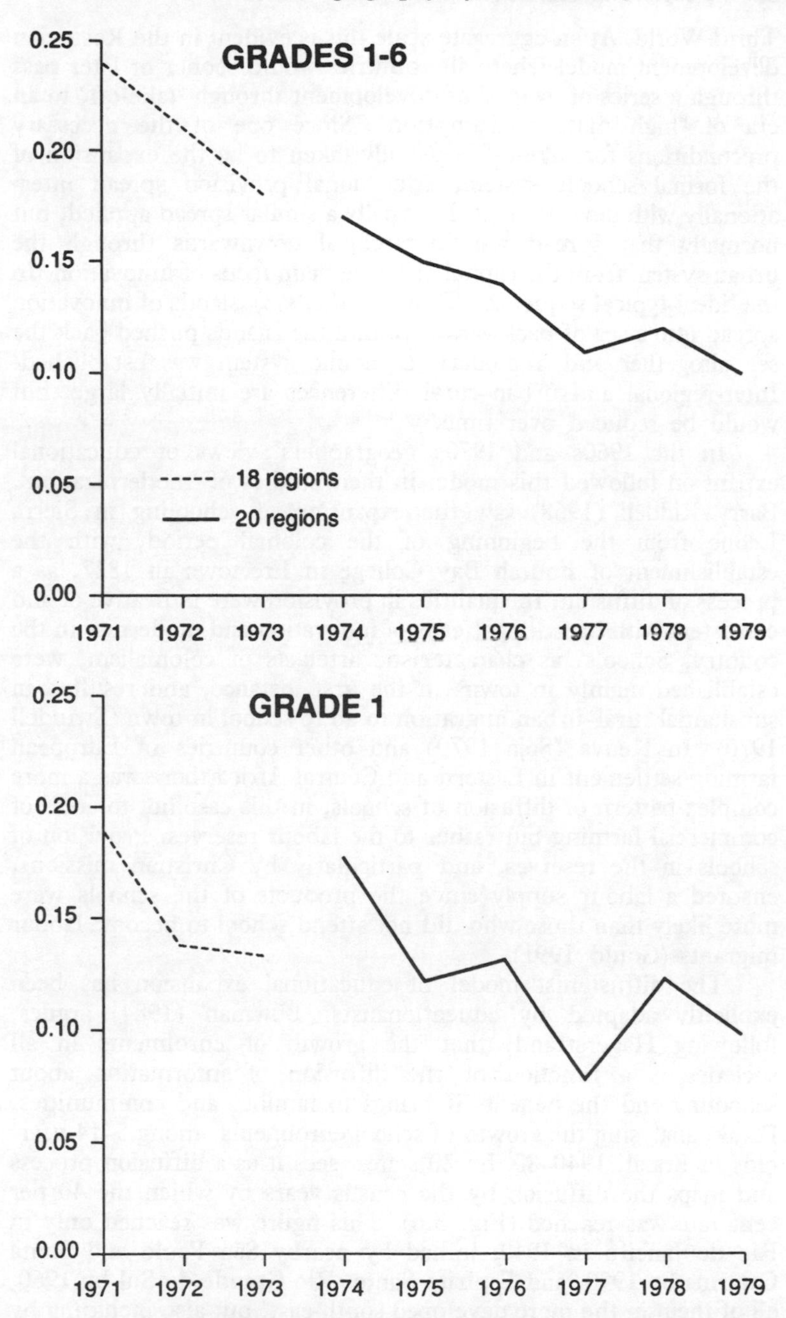

Fig. 3.5 Gini coefficient for Grades 1–6 and Grade 1 enrolment rates by province: Papua New Guinea 1971–79

Third World. At an aggregate scale this is evident in the Rostovian development model where all countries would sooner or later pass through a series of 'stages' of development through 'take-off' to an era of 'high mass consumption'. Since one of the necessary preconditions for take-off is usually taken to be the expansion of the formal schools system, educational provision spread internationally with development. Internally a similar spread applied, but normally that spread was hierarchical downwards through the urban system from the capital city, the main focus of innovation. In the 'ideal-typical sequence' (Taaffe *et al.* 1963) islands of innovation spread into a sea of backwardness until the islands pushed back the sea altogether and a modern economic system was established. Inter-regional and urban–rural differences are initially large, but would be reduced over time.

In the 1960s and 1970s geographers' views of educational expansion followed this model in their studies of 'modernization'. Barry Riddell (1968) saw the expansion of schooling in Sierra Leone from the beginning of the colonial period, with the establishment of Fourah Bay College in Freetown in 1827, as a process of diffusion. Inequalities in provision were indicative of and consistent with broader patterns of innovation and modernity in the country. Schools, as characteristic artefacts of colonialism, were established mainly in towns in the first instance, and resulted in substantial rural–urban migration to go to school in town (Swindell 1970). In Kenya (Soja 1970) and other countries of European farming settlement in Eastern and Central Africa there was a more complex pattern of diffusion of schools, in this case not to areas of commercial farming but rather to the labour reserves. Provision of schools in the reserves, and particularly by Christian missions, ensured a labour supply since the products of the schools were more likely than those who did not attend school to become labour migrants (Gould 1991).

The diffusionist model of educational expansion has been explicitly adapted by educationalists. Bowman (1984) argues, following Hagerstrand, that the growth of enrolments in all societies is a function of the diffusion of information about schooling and the benefits it brings to families and communities. Plank, analysing the growth of school enrolments among 5–14-year-olds in Brazil, 1940–80, by 20 states, sees it as a diffusion process and maps the diffusion by the census years by which the 40 per cent rate was reached (Fig. 3.6). This figure was reached only in Rio de Janeiro in 1940, joined by nearby São Paulo and Santa Catarina by 1950, and Espirito Santo, Rio Grande do Sul by 1960, all of them in the more developed south-east, but also including by then Mato Grosso and Para. By 1980 only three states (Maranhao, Piaui and Alagoas), all of them in the impoverished north-east, had

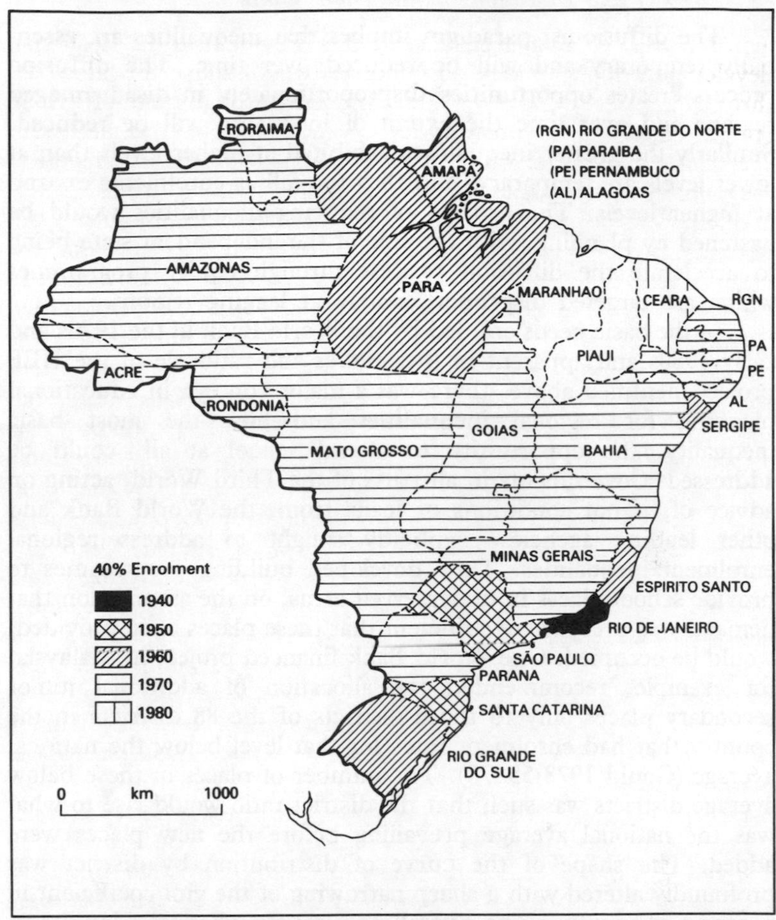

Fig. 3.6 Year by which 40 per cent enrolment had been achieved by state: Brazil (from Plank 1987:364)

not reached 40 per cent enrolment of that age group in any census year. Bahia, also in the north-east, was also below 40 per cent in 1980, but it had been 42 per cent in 1970. Plank concludes that there was not a simple spatial or temporal diffusion, but that the expansion was closely associated with broader features of socio-economic change in the country, notably urbanization and changes in occupational structure that affected the demand for educated workers. Throughout the period, the ranks of the 20 states remained broadly similar, and all states of the country were gradually integrated into the national system. At a different spatial scale, though, inter-municipality differences within Brazilian states remain very substantial and greater than inter-state differences (Brooke, 1992).

The diffusionist paradigm implies that inequalities are essentially temporary and will be reduced over time. The diffusion process creates opportunities disproportionately in disadvantaged regions and over time the extent of inequality will be reduced. Similarly the greater inequalities exhibited at higher levels than at lower levels are temporary, and they too fall as enrolments expand at higher levels. The process of reducing inequalities would be hastened by planning, with the role of the independent state being to accelerate the diffusion process through equity programmes which are targeted disproportionately at lagging regions.

In the basic needs era, led by the World Bank in the 1970s and early 1980s and apparent in the studies and rationale of the IIEP project identified above, there was a major concern in educational planning for regional inequalities and how the most basic inequality, the opportunity to attend school at all, could be addressed. Governments in all parts of the Third World, acting on advice of, if not conditions of loans from, the World Bank and other lending agencies, explicitly sought to address regional enrolment inequalities. They developed building programmes to provide school places in underserved areas, on the assumption that demand was everywhere sufficient that these places once provided, would be occupied. One World Bank financed project in Malaysia, for example, recommended the allocation of additional junior secondary places only to those districts of the 88 districts in the country that had enrolment rates for that level below the national average (Gould 1978:55–78). The number of places in these below average districts was such that the district ratio would rise to what was the national average prevailing before the new places were added. The shape of the curve of distribution by district was profoundly altered with a sharp narrowing of the gini coefficient in a period of rapidly rising overall enrolments where the long-term objective was universal enrolment at that level.

The assumptions of the diffusionist model have been dominant in educational planning throughout the main period of enrolment expansion in the Third World. The independent state was seen to be the chief agent for the diffusion, promoting equity in education as in other social sectors.

The underdevelopment paradigm

Despite the massive expansions, planned or unplanned, spatial disparities in enrolment levels have persisted. Only where there is compulsory universal enrolment are quantitative disparities eliminated, and even then, as in Western Europe and North America, qualitative disparities persist to maintain a significantly variable geography of educational opportunity. In Third World countries, however, despite the objectives of government, large quantitative disparities persist. The extent of disparity at any given level may fall over time, particularly where there are planned expansions with the objective of reducing inequalities. Where direct involvement by governments in planning has been weak and expansion largely dictated by the pace and spatial pattern of demand, as in Brazil, then the range of enrolment rates amongst regions may not necessarily fall and particularly in periods when overall expansion is small. Plank's data show for Brazil between 1970 and 1980 a small improvement in the national 5–14 enrolment rate from 56 to 57.5 per cent (see above). In that period, however, the rate fell in seven states, including three of the five poorest states, all in the north-east. It rose from 75 to 77 per cent in Rio de Janeiro and from 66 to 70 per cent in Sao Paulo, the states with the highest rates, but fell from 40 to 38 per cent in Alagoas and from 40 to 36 per cent in Maranhao, the two lowest states in enrolment levels. The diffusion process seems to have gone into reverse! Growing inter-state disparities in enrolment are in part associated with growing rural–urban differences in education as in other aspects of Brazilian life. Wood and Carvalho explore 'The demography of inequality in Brazil' to show how for most income groups there were in the same period, 1970 to 1980, widening disparities between rural and urban areas in the proportions of women with more than four years of schooling (Wood and Carvalho 1988:171). They argue more generally that there were structural forces, internally and externally, operating to maintain a range of geographical (inter-state and rural–urban) and social inequalities in Brazil, and that these forces are part of the broader processes of creating and maintaining inequalities in capitalist societies.

In general terms, the diffusionist paradigm in development studies has receded in its importance for scholars in favour of the underdevelopment or disequilibrium paradigm. As a country develops, all parts of it do not develop equally and there are systematic processes of underdevelopment that allow some areas to develop faster than, and often at the expense of, other areas. Regions are part of an integrated national system, and development in one part of that system will affect the possibilities for the pace and type of development in all other parts of the system. Regions where demand for schooling is high are generally those where income is highest and which have greatest political clout. They will continue to be allocated a disproportionately large share of resources for education by government. Furthermore, individual families will be more able to allocate their own resources to keeping children in school. Even where there is rapid overall expansion in all regions at the basic level, access to higher levels will remain in practice biased in favour of students from the most advantaged areas. There is a cumulative process of comparative advantage, in Myrdal's terms, that extends to the education system, allocating new resources (teachers, materials) disproportionately to already advantaged areas, and attracting resources (e.g. existing teachers in post) from disadvantaged areas. Regional disparities are to be expected in a market system, for the stronger demand in the richer areas will be reinforced by the spatial pattern of supply of public and private schools.

Regional imbalances are paralleled by and systematically associated with 'urban bias' (Lipton 1977), the systematic bias of the market and the state in Third World countries in favour of urban development rather than, indeed often at the expense of, rural areas. Everywhere urban enrolment rates at all levels are higher than in rural areas. Urban schools are better built, have more highly qualified staff and better facilities. The quantitative disparity is likely to be exacerbated by a qualitative disparity. Since, as we have seen (Chapter 1), the education system tends to reflect modern, Western values, values that are most evident in urban areas, there is a logical consistency between urban bias in provision and urban bias in demand. There tends to be a preference in opportunities to progress in the schools for urban areas where these urban values and opportunities can be lived out. Rural areas lose their educated people through migration in an internal 'brain drain' (see Chapter 7). The demand for more and better rural schools is further reduced as the 'spread' effects, that push provision out to rural and backward areas, give way to 'backwash' effects of the education system itself pulling resources and people into towns and richer rural areas, as well as to the backwash effects of the overall urban bias in economic and social development.

Reducing inequality

In the face of the 'natural' tendency towards regional and rural–urban inequality, governments of Third World countries have pursued a range of possible strategies. Some have sought to prevent widening disparities, either by positive allocations or more negatively by limiting expansion in some areas; others have taken a more passive view in the face of variations in demand. In federal systems there is a wide range of national approaches to the need for, as well as the mechanisms of, equalization strategies through financial allocation to component states (Hinchliffe 1989). In

Brazil, despite large spatial inequalities, there is relatively little emphasis in redistribution in inter-state budget allocations in higher education. Allocations are primarily based on the 'neutral' criterion of population. India has rather similar presumptions at that level, but the extent of inter-state inequality in provision and enrolment is rather less, and there is some allowance for each state's income and local tax-generating capacity. In Nigeria, by contrast, there have been considerable changes in allocation criteria in recent years, but all have had a great emphasis on differential allocations towards equalization of regional differentials in higher education.

In centralized, unitary systems there is normally explicit concern for regional inequalities, but this need not be directly reflected in public allocations. In Kenya, for example, central government has provided basic provision but this is complemented by a shadow system of private schools, in many respects outside the control of the Ministry of Education – the *harambee* schools. These were financed and built through self-help efforts in a major nationwide movement after independence to overcome the facilities shortages inherited from the colonial period (there are also *harambee* clinics, cattle dips, community halls, etc.). Most primary *hurambee* schools have subsequently been integrated into the government system, but at the secondary level there are many *harambee* schools ('unaided'), and others where some classes and teachers are maintained by government, others by the community ('assisted'), while the largest single group of enrolments is in the maintained sector. Well over half the total enrolment for boys and girls is in the unaided or assisted sectors in forms I–IV, the basic secondary level (Table 3.2), but there are very few such schools at the higher secondary level. Unaided or assisted schools are disproportionately in towns and better-off rural areas, such as Central and Rift Valley provinces where families have the income to pay the high fees that are needed (those who go to *harambee* schools are often those who have failed to be offered the much cheaper place in a government school, where teachers are generally much better qualified and the learning environment generally very much better). Girls comprise a larger proportion of total enrolments in assisted and unaided schools than in maintained schools, one reflection of these higher incomes.

As far as the Government of Kenya is concerned the main disadvantage of such a large private sector is that it will raise the supply of school leavers seeking a diminishing number of jobs, thus exacerbating overt urban unemployment. In addition it will increase the extent of inter-regional disparities in enrolment. These disadvantages are outweighed, however, by these school leavers being able to help satisfy demand in the richest and most politically

Table 3.2 Enrolment in secondary school by level, sex and type of school: Kenya, 1987 and 1988

	1987			1988		
	Maintained	Assisted	Unaided	Maintained	Assisted	Unaided
Forms I–IV (Secondary)	214 779	172 536	91 021	219 343	180 676	86 021
Forms V–VI (Higher Secondary)	36 362	1 554	6 009	44 236	2 287	7 629
Boys	155 930	97 164	54 950	162 837	100 643	54 521
Girls	95 211	76 926	42 080	100 742	82 320	39 129

Source: Kenya Government 1989c:151.

influential areas, areas where the off-farm urban and rural informal sector is sufficiently vigorous to absorb many poorly certificated school leavers. Increasing regional disparity for Kenya is an inevitable consequence of allowing spontaneous community demand to be met.

At the other extreme many countries have sought to directly confront regional disparities through centralized planning controls. In a few countries schools were discouraged or prevented in areas where their establishment would greatly widen existing inequalities. Tanzania, for example, a country greatly concerned to promote equality of opportunity after the Arusha Declaration and the publication of *Education for self reliance* in 1967, stoutly resisted the establishment of private secondary schools even though the proportion of the age cohort finding a place in government secondary schools was one of the lowest in the world. This gave rise to a popular demand for private schooling that the government was reluctant to satisfy. The demand was strongest in Kilimanjaro Region, the richest region in the country and already supplying a disproportionately high number of secondary school students (Samoff 1987). By the late 1980s, however, the centrally imposed prohibition on private secondary schools had been of nccessity relaxed in the face of public disquiet in that region and elsewhere.

Governments have generally recognized the inevitability of regional disparities, but have sought positively to address them through planning procedures in the Ministry of Education. These procedures have tended to support a positive discrimination towards lagging regions in the allocation of the inputs over which the Ministry has control. The planning criteria cited above in the case of Malaysia with the allocation of inputs only to those districts below the national average and up to sufficient to reach that average is a vigorous example of what can be proposed. A gentler approach to the narrowing of disparities is more familiar, partly for political reasons (all regions want at least some share of new resources, however small) and partly to support the national upward trend in enrolments or in teacher supply. This will typically involve allocations to all regions, but more to some than to others to narrow the gap and lower the gini coefficient. Where secondary school places were in very short supply, as in the 1960s in many African countries, and the majority of existing schools were boarding schools, access to them by the most able students was achieved by a centralized national admissions system (Gould 1974, 1975). In other cases, for example in Federal Government secondary schools in Nigeria, regional balance in admissions was guaranteed by formal regional enrolment quotas.

As schools systems have expanded, central control has become increasingly difficult and, even in states that are otherwise highly centralized in their administrative structure (e.g. Indonesia), government is either unwilling or unable to intervene strongly to reduce quantitative inequalities. A general trend in Third World countries in the 1970s, 1980s and into the 1990s has been towards decentralization and participatory democracy, often a condition of externally financed programmes rather than as a result of a great outpouring of people power, and positive restructuring introduced by centralized national governments. One fairly spontaneous case, however, was Papua New Guinea, which established 20 provinces each with its own Ministry of Education with its own planning powers, in a country that had experienced very considerable regional disparity in enrolment (Bray 1984). The effect of decentralization in education was to narrow disparities slightly, with rapid expansions in the formerly lagging provinces of the Highlands, but much less than National Government would have wished. However, it is clear that there was greater equality in the social sectors generally (health, housing, infrastructure) over which provinces had some control than in economic activity over which they had little control (Berry and Jackson 1981).

In 1989 in Ghana, planning responsibilities in basic education (6 years of primary and 3 years of junior secondary) were transferred from a previously highly centralized Ministry of Education to 110 newly established districts. These districts are now supported financially to meet national minimum guidelines for centrally established enrolment levels, teacher supply, etc., but many districts already exceed these minimum targets. A newly established database of inter-district diagnostic indices identifies a wide range of qualitative and quantitative features of the system at all levels, as has been evident in Ghana since modern education was introduced (Gould 1990b; Hunter 1964; see also Fig. 3.2). There is a broad difference between north and south. Relatively high levels of enrolment occur in the richer south, where population densities and per capita demand are higher than in the savanna regions of the north, with low overall densities and relatively weak demand for schooling. Decentralization, as advocated for Africa as a whole by the World Bank, may help solve some problems (e.g. in teacher performance, in local accountability of schools), but it is not likely to facilitate any significant reduction of inter-district inequalities. The richer and already better served districts will be more able and willing to raise their own revenues to supplement nationally distributed resources, and the educational gap between the southern and northern districts is likely to widen without strong enforcement of centrally agreed minimum levels of provision, e.g. for minimum enrolment levels or proportions of qualified teachers, or by setting district maximum or minimum targets in nationally sanctioned district education plans.

Regional inequality: challenge or opportunity?

Perhaps the most interesting views on regional disparities in education are associated with the Swiss educationalist Pierre Furter (1980, 1983). He argues that a lot of formal education, in the developed world as in the Third World, has had a homogenizing effect on peoples. It has stifled cultural differences within states and has been designed to support national objectives, political and economic, that have undermined regional self-sufficiency and distinctiveness, especially in the realm of culture. His is a plea for regional distinctiveness in education. This is not a distinctiveness that denies an opportunity to go to school or to progress in school to some children and not to others, but a distinctiveness that uses and emphasizes local characteristics of society and the environment and permits local school systems to develop their own curriculum within national norms, and in local languages. Furter's own national experience in Switzerland, with its highly decentralized education system is the model for his views, for it is founded on local needs and capabilities rather than a national strait-jacket. He explores the difficult objective of reducing regional disparities and inequalities 'in a way that can respect regional diversity' (Furter 1980:51). For him regional disparities are challenges to be accommodated rather than inefficiencies or inequalities to be eliminated. In this he develops an argument similar to that of Bjorn Hettne (1990) that development into the 1990s needs to be sensitive to ethnic aspirations at the local scale. 'Another development', that is based on neo-populist and locally appropriate models, needs to be integrated into national planning stategies in education.

Chapter 4

The local scale: the school and the community

Having dealt in the aggregate at the global, national and regional scales in the previous chapters, consideration must now be directed to the local scale. This will examine the individual school and its relationship with the community in its immediate catchment area and, more directly, with the children it serves. Education is most commonly delivered in fixed site buildings, and takes the institutional form of 'schooling', more narrowly defined than 'education'. Schools are inevitably physically within local communities and linked to the economic and social life of the community in several respects. They are familiar physical features: characteristic classroom blocks in a variety of building styles – both local and imported – often with a large play area, especially for rural schools. They are often identified by a prominent signpost. Their physical location, often near the centre of a village or other settlement, seems to be symbolic of their broader social and economic significance for the community.

This chapter considers the school/community relationships in two broad sections. The first half of the chapter examines the physical aspects of the location of schools and access of students to them. It starts with the premiss that a school is a particular type of service facility, akin to a health centre or a cattle dip or a water supply point, provided at a fixed site to which the consumers come to use the facility, leaving their home to attend a central facility. It is possible to deliver education in the home through distance-learning techniques, i.e. radio and TV or correspondence courses. These are certainly important in Third World countries, and especially for adult literacy programmes or family planning media campaigns, but for young children the socialization and discipline of formal schooling is an important component of the whole learning process. The particular issues raised by the nature of the journey to school give rise to a range of educational and social planning issues that affect the type of school provision – the size of school, its distance from other schools, its relation to these other components of the social and economic infrastructure. These issues are strongly associated with broader questions of geography and planning at the local scale.

Mere provision of a school, however, is not in itself critical to the relationship between the school and the community it serves. Analysis must therefore go beyond the locational planning issues addressed in the first half of the chapter to explore how the presence of a school can affect the economy and social and cultural life of the community. In which respects and to what extent can a school be a focus for community life and a catalyst for local economic change? By contrast, in which respects and to what extent can and does a school reflect the wider constraints and tensions within the community and further act to sustain or even exacerbate these constraints and tensions? Issues such as these are introduced in the second half of the chapter. They are discussed with particular reference to some of the locational planning questions raised in the first half, and also to issues of decentralization and local control raised towards the end of Chapter 3.

Schools and the geography of social provision

It is expected that there should be a close correlation between the distribution of schools and the distribution of the population 'at risk': the children in the relevant age groups that the schools serve. Where there is a mismatch at the aggregate scale, the disparities that were considered in the two previous chapters become apparent through comparison of enrolment rates. These aggregate disparities arise because of disparities at the local level. Some schools may be inappropriately located, so that some children live too far from a school; or else there may simply not be sufficient places available within an area even though children live near a school. Of course it would be ideal to have a school next door to each home, just as it would be ideal to have a clinic next door to each household, but the number of schools needs to be rationed to ensure a system that can operate within acceptable cost parameters. This is the essential problem in the geography of all social provision, the distribution of any facility: how to strike a balance between on the one hand the desirability of maximizing access to that facility and thereby maximizing *equity*, and on the other hand the desirability of minimizing the cost of providing the service and thereby maximizing *efficiency*.

Some services (e.g. clean water, mail, electricity) are ideally provided directly to consumers in their homes, and elaborate and often costly delivery systems have been developed to facilitate these. In Third World countries the sheer cost of direct public provision of even these is normally prohibitive, particularly in rural areas, but also in urban areas where there are substantial differences between neighbourhoods according to ability to pay. The tendency since the mid-1980s in Third World countries has been to move from public 'free' provision of such services towards a user-pays principle, whether in services that are public monopolies or in private sector provision (Roth 1987).

Schools are therefore similar to clinics or other community facilities in that there is a presumption of personal attendance to use the service provided. Users attend school in order to have access to education. Reducing spatial inequalities in school attendance therefore crucially depends on improving physical access, on having more schools, and on removing economic and social barriers to admission to existing schools. Obviously simple expansion of the number of schools and places in existing schools will increase access in an aggregate sense, but the location of these new or expanded facilities will be critical to the spatial pattern of access. Appropriate locations will ensure that previously unserved or underserved communities have real access, in a physical sense, and not merely improved general access, as where the number of places in school in a region rises but these new places are provided in areas where there is already a school.

Equity strategies require explicit concern for the distributions of schools. As has been discussed in Chapter 3, there are normally differential levels of enrolment within Third World countries at a regional scale, with urban areas having higher levels of provision than rural areas, and commercially prosperous rural regions near the main towns or main communication routes higher levels than poorer or more remote regions. When the expansion of the schools system is considered as a hierarchical diffusion process, with expansion from the capital through the urban hierarchy downwards to the whole country, as in the case of Brazil (see Chapter 3), then the extent of the local disparity in access to schools can be seen as a function of time. Sooner or later – and most probably later rather than sooner – schools will spread to the remote rural communities, with the assumed operation of a 'trickle-down' effect, as with other aspects of innovation and modernization.

However, this trickle-down will not necessarily happen spontaneously. Indeed much of the evidence of the last few decades seems to imply that there are strong forces constraining the spread of innovations, to the extent that in many circumstances spatial inequalities may be widened rather than reduced in the early stage of development. This is as true at the local scale as we have seen it to be at the national scale. If there is to be a strong trickle-down of schools into rural areas then it may need to be stimulated through direct intervention in the planning process. The technical possibilities of implementing planning procedures to ensure a more rapid diffusion of schools are not sufficient to ensure that the diffusion will necessarily occur. They need to be underpinned by a political commitment to an equity strategy that seeks to promote equality of opportunity, at least for initial access to school. Most Third World countries are committed in general terms to such an objective and have therefore sought, with mixed success, to ensure a more appropriate distribution of schools through better planning at the local scale. Achievement of that objective has met with mixed success.

That national commitment was strongly supported in the late 1970s and into the 1980s by an international commitment to equity-driven planning and, particularly in the 'basic needs' policies of the World Bank at that time, given immense priority by its then President. Expansion policies in the poorest countries were to give priority to projects which were directed to poverty alleviation, and therefore to the poorest people in the poorest areas. This meant priority to more productive food crops rather than to cash crops; to rural water and sanitation improvements; to rural primary health care (PHC) rather than sophisticated urban hospitals; and to rural primary schools rather than (but not always instead of) national universities. The 1974 Education Sector Paper of the World Bank stated: 'The appropriate location of educational facilities is a simple but effective instrument, particularly for lower levels of education where physical proximity is a major factor determining enrolments' (World Bank 1974:34). Such a statement seems so obvious that it is hardly worth making, yet it does recognize that an appropriate distribution of schools will not always occur spontaneously, and may need to be planned.

This need was operationalized within the World Bank through a set of techniques and procedures termed *school location planning* (Gould 1978). These were designed to assist the preparation and appraisal of projects for enrolment expansion to be financed by the World Bank as part of the broader basic needs approach. They had been preceded by an initiative of the UNESCO-based International Institute for Educational Planning in what it called 'school mapping', much derived from the French educational planning techniques of *la carte scolaire* (Hallak 1977). Within UNESCO and IIEP the approaches have evolved to be known as 'educational micro-planning' (UNESCO 1983). There is nothing particularly new or innovative in a technical sense about these methods. Those responsible for the management of schools systems had always had to allocate new schools and other inputs (teachers, equipment and other learning materials) into them, but the more systematic methods implicit in these formal approaches presumed that there might be a problem of misallocation and increasing inequality at the local level, for technical as well as political reasons. They sought to establish structures and apply allocation criteria that would allow projects for educational expansion to achieve equity objectives more effectively, both more rapidly and with greater cost efficiency. At their core, though not always made explicit, are the twin concepts of 'threshold' and 'range', familiar to geographers and planners and strongly dependent upon the normative assumptions of central place theory, i.e. that consumers act 'rationally' in a spatial sense in that they minimize the distances they travel to use the facility. There is a recognition of schools as providing a service and their need to be located near to the consumers of that service. It is a demand-based, i.e. population-based, approach to public and, to a lesser extent, private provision.

The journey to school

The journey to school has two characteristic features. In the first place it is regular. It occurs in the morning and in the evening each day the school is in session, and for many children in many schools at lunchtime too. In the second place, it is done by young children from about 5 years in most countries. For both these reasons the journey will be very short, in distance as well as in time. There must therefore be a close relationship between the distribution of schools and the distribution of population. Empirical data on the precise nature of the distance decay function for schools were derived in the case studies which provided the basis for the IIEP methodology (Table 4.1).

In a range of Third World countries, including Costa Rica, Nepal, Iran, Lebanon and four African countries, a consistent pattern of short travel distances for primary school children was found. Typically over 90 per cent of children have a journey of less than 3 km. While most of the studies measured the journey in distance, the case study in Nepal, in a sharply mountainous region where linear distance is likely to be seriously misleading, the measurement was in time. In that case 90 per cent of primary school students had journeys of less than 45 minutes, but this proportion fell to 75 per cent for secondary level students. In some of the case studies comparable data are not available and are omitted from the table. In the Iran case study, for example, undertaken in a very sparsely populated area with a nucleated settlement pattern, most students attended school in their own village. In only 15 schools of the 146 in the Chahroud Region were

Table 4.1 Journey to primary school: proportions at a distance or time period

Distance									
Km		1	2	3	4	5	6	7	8
Miles		½	1	2	3	4	5		
Minutes		15	30	45	60	75	90	120	
Uganda	– P1	35.0	28.6	22.0	10.9	2.4	1.2	0.7	–
	– P6	21.1	21.1	24.5	17.1	8.4	3.4	1.0	0.3
Nepal[1]	– 1st Level	61.3	28.7	8.7	0.6	0.7→			
	– 2nd Level	34.7	24.1	13.5	14.3	8.5	4.9		
Lebanon	– Public	95.3	1.7	0.5	2.4[illegible]				
	– Private	90.5	5.6	1.4	2[illegible]				
Morocco	– Primary	27.9	43.5	18.7	7.8	2.1→			
Costa Rica	– Boys	83.1	16.9 →						
	– Girls	86.1	13.9 →						
Ivory Coast	– Sikensi[2]	23.9	29.0	37.0	2.2	6.5	[illegible]		
	– Dabakala[2]	0.4	11.6	23.2	22.8	33.6	[illegible]		
Algeria	– Primary	81.0	9.0	4.5	1.3	[illegible]			

Source: Gould 1978: 87.

[1] It was assumed in the case of Nepal that it took children 30 …

[2] Comprising only those pupils who do not attend school i…

→ Indicates an open-ended category.

ese travelled less than

can be related to
ment distribution (see
ral picture of a sharp
e *range* of a school, to
be explicitly recognized
ler for early grades of
children, and larger for
young adults who are
ces in a shorter period of
the range are not only a
distance regularly. They
ending children to school
a regular and substantial
the household, whether in
ic chores, or looking after
tivities, the time spent in
school and going to school
weighed against these other
expectations and commitments. Long distances and long journey times raise the opportunity cost of sending a child to school. Where these are long, the perceived benefits of not sending a child to school may be sufficiently high to persuade them and their parents that initial non-attendance or later drop-out is preferable.

Furthermore there are good educational reasons for minimizing the distance travelled. Where distances are excessive the children are more likely to arrive in school tired and less receptive to learning. They may doze for long periods, especially if they are undernourished. They are more likely therefore to perform less well and to become drop-outs for educational rather than strictly distance reasons, though these remain important. When they are slightly sick, or when it rains, the longer distance is a factor in infrequent attendance and will raise the propensity for drop-out. To reduce the mean journey to school will be likely to raise educational achievement and reduce drop-out rates, both important objectives of educational planning.

The IIEP case studies were done in predominantly rural areas of poor countries where the overwhelming majority of students travelled to school on foot. In the Moroccan case, for example, 96 per cent of pupils walked; only 3 per cent came by bicycle and only 1 per cent by public transport. Travel on foot is likely to remain characteristic of school-going for most children in the Third World. However, in some cases and for particular types of school, the tyranny of distance can be overcome and the range extended. Extending the range can be achieved either by altering the number of journeys and their regularity, or by altering the means of transport. Where the journey to school can be less regular it can be longer. Instead of commuting every day, there can be weekly, termly or even annual movement from home to school to live in accommodation in or near the school within a normal daily 'walking' range of the school attended.

The classic solution to the distance problem in the early period of modern educational development was the boarding school, often the mission boarding school. In British colonies this was modelled on the English upper middle-class boarding school. The boarding school allowed access to secondary school by children whose homes were far beyond daily access to such a school, and this was the dominant pattern in the colonial years when there were so few secondary schools. Boarding schools were the characteristic type of secondary school in many African countries in the colonial period and immediately after independence, and were important in ensuring access by some children from remote rural areas. Criteria for admission were administered centrally (or at least regionally), on the basis of a national primary leaving examination, in order to improve access by children from rural areas (Gould 1974). The boarding school also had the added advantage for nation building that it allowed mixing of pupils from different ethnic or linguistic backgrounds in the expectation of better mutual understanding by often traditionally antagonistic ethnic groups in the creation of what was, at the time, a small national élite. The secondary boarding school broke the direct link between home and school, and encouraged socialization of its students in the national ethos. However, it also alienated students in many respects from traditional culture and a daily life of poverty and struggle that was familiar to their families. Though they were initially necessary for ensuring a spread of participation, boarding schools came to be viewed as élite schools. Boarding became less and less necessary for strictly access reasons as more children were living within daily access. There was a rapid growth of rural day-secondary schools with the expansions of the 1960s and 1970s. In most countries many of the old boarding schools have remained as prestigious institutions, increasingly the preserve of children of the élite, but still heavily subsidized from the public purse.

The principal disadvantage of the boarding school, however, is a financial one: they are expensive to run and, where the boarding costs are borne mainly or even entirely by the Ministry of Education, they were soon perceived to be an expensive and unnecessary luxury. In increasingly hard-pressed Third World countries and under pressure from donor agencies, the emphasis in secondary education (and indeed in universities) has been towards day provision, with the assumption that where leaving home was necessary the costs of accommodation would be borne directly by the student and not by the school. In some cases this has required existing boarding schools to be converted to day schools. Accommodation has been privatized in practice or, less commonly, closed, and there has been a complete restructuring of the catchment areas of schools. In Ghana from 1988 a major programme of 'deboardingization' of senior secondary schools was introduced as part of major reforms in education, moving towards 'cost recovery', in practice transferring the costs of boarding to the consumers. Similar assumptions have been applied in the higher education sector where costs of tuition (still subsidized) are separated from the costs of accommodation. The changes brought sharply increased costs to individual students. Such moves clearly have social as well as spatial implications, for they tend to favour better-off students from urban backgrounds against the poor students from remote areas. But they also strengthen the school/community links.

The range of a school can also be extended with improved availability of and access to public or private transport, to allow greater distances to be covered in the same period of time taken to walk to school. In developed countries, where transport is in any case more widely available, this may mean a legal requirement for the state to provide transport to children living at a distance from a school, but in poor countries with poorly developed transport infrastructure this is not always a feasible option in rural areas. Only along major existing transport routes with a good system of buses or shared taxis (e.g. 'jeepneys' in Philippines, 'trotros' in Ghana or 'mutatus' in Kenya) can vehicle transportation make a significant contribution. A more likely solution is offered by bicycle transport, and especially in the more prosperous countries, though the use of the bicycle as a cheap and relatively accessible means of private transport has important cultural variants. In tropical Africa, for example, bicycles are relatively uncommon and there is some resistance to their use even among school children. In Indonesia, by contrast and probably associated in part with its Dutch colonial heritage, there is widespread use of bicycles by school children. Most rural secondary schools have large bike sheds and can generate major traffic bottlenecks before the schools open and after they close each day.

In towns, however, the transport option is most readily available, though ironically where it is least needed, for distances from home to school can be much less and the whole town can be within access to schools with a small range. Furthermore, since there are higher rates of private car ownership than in rural areas, more children can be taken by their parents to school by car and catchment areas can be large, often city-wide.

The transport option is available to some people in some areas to improve access. As income levels rise and transportation systems improve that option will be increasingly available. However, it is in the areas of poorest access and lowest enrolment rates that this option is least feasible, and educational planners are obliged to continue to develop patterns of provision on the assumption of small and relatively inflexible catchment areas with short journeys to school. These bring inevitable consequences for the size of schools.

Population threshold and the size of school

Given these small catchment areas, even for secondary schools, the main condition for maximizing access to education is the need to have many schools that are well distributed throughout an area such that no child lives further than perhaps 5 km from a school. This is not normally possible, even in the best served countries, for the internal efficiency of the schools requires that they should normally be relatively large. Only in areas of very high population density, such as in towns, will there be sufficient potential pupils within the range to allow a school to operate at an acceptable level of efficiency. A problem common to planning all facilities is to identify the appropriate population threshold, that population required within the range to allow the facility to operate at an acceptable level of use. While it is a relatively complex matter to identify the threshold population for a water outlet or dispensary, since use of these facilities can be highly variable within populations, in the case of schools it seems more straightforward since there is a relatively easily identifiable 'at risk' group for whom a regular service needs to be provided, and that group is normally a known proportion of the total population. Census data will normally provide a breakdown of population data by age for the notional educational groups. In most African countries, for example, where there are high rates of population growth and high proportions of that population in the school age groups, roughly one-sixth of the population, or between 16 and 18 per cent, is aged in the primary school age group. The threshold population of a school varies from level to level, being lowest for primary schools and rising for each level of the education hierarchy. This is particularly the case where there is not universal enrolment at any level and the proportion of the relevant age-group actually in school falls as the level rises. Even though there may be 95 per cent enrolment or more at primary school level and perhaps 75 per cent at junior secondary, this proportion may typically fall to below 50 per cent and in many cases much lower than that for secondary levels, further raising the total population needed to generate the mumber of pupils required.

A school is assumed to provide education for a complete cycle – a primary school with a six-grade first cycle will normally be expected to have six classes and six teachers, and perhaps also a headteacher in addition. For secondary schools with specialist subject teachers there may need to be even more than six teachers in a school, even where one teacher may teach two or even more subjects. However, teachers are by far the most expensive input into a school. Their salaries normally account for far in excess of 90 per cent of the costs of running the school, a proportion that can rise to almost 100 per cent in the poorest countries where little or no equipment or learning materials are provided. The most efficient schools must therefore have high pupil/teacher ratios if costs per pupil are to be minimized. Typically the national norm for pupil/teacher ratios is around 40:1 for primary schools and around 30:1 for secondary schools. Thus for an ideal primary school with six teachers and a pupil/teacher ratio of 40:1, these would be expected to be 240 pupils (assuming no drop-outs that would mean less than 40 pupils in the upper levels of the cycle where there are 40 pupils at the initial level). Approximately 1400 people would need to be the threshold population for a primary school where the school age group was one-sixth of the total population and there was 100 per cent enrolment. The threshold populations for higher levels of the education system, where schools may need to be even larger and enrolment rates more likely to be lower, will be substantially higher than 1400.

Will that threshold population be reached within the fairly small range of the school? With a range of 5 km the area of the daily catchment of a school is approximately 80 km^2 (assuming a circular area), and the threshold population density will therefore be 175 per km^2 for a population of 1400, assuming a population evenly distributed throughout the area. Much of course will depend on the settlement distribution. Where the population is nucleated, i.e. living in recognizable villages, the size of the individual settlement will be critical; where the population is dispersed and well scattered across the region then the mean density will be critical. The critical importance of settlement pattern is identified in Fig. 4.1 in the contrast between area A and area B. In both areas there are 10 000 people. In A they all live in the three villages, but in B they are well distributed throughout the area. The schools and their catchment areas are similar, but in A the whole population is within their range and in B it is not. In B the demand surface (XY) is evenly spread, but it is highly concentrated in A in the three villages in which the three schools are located. Clearly in urban

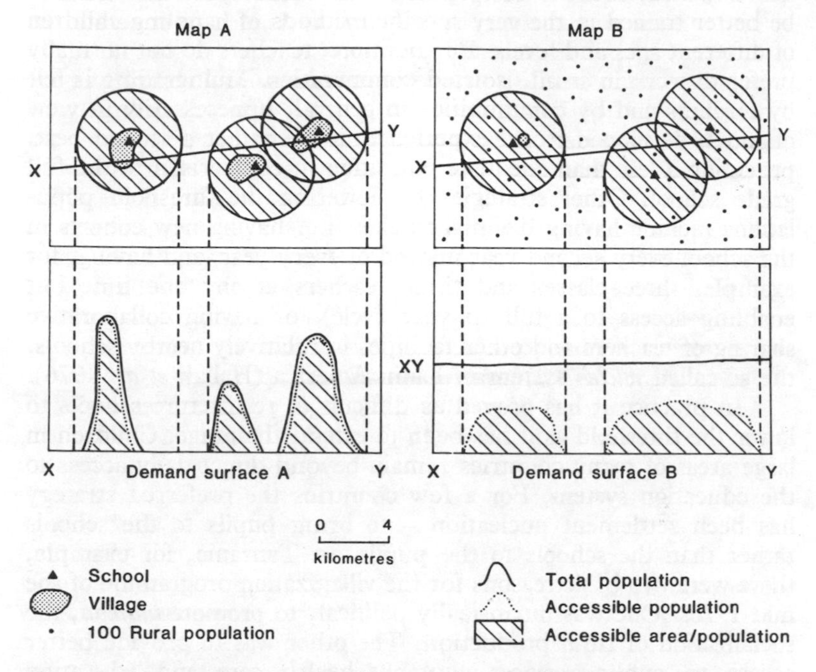

Fig. 4.1 Accessibility and local population distribution (from Gould 1978:91)

areas and in high density rural areas access is not likely to be denied where there are schools, but the problems of access will be most apparent in areas of low overall population density, and especially where there is dispersed settlement.

Threshold populations tend to be high, and generally higher than the mean national rural population densities (Gould 1982a). Since, as we have seen, there are serious constraints preventing the extension of the range, one means to improving access might be to lower the threshold, to alter the internal workings of the schools to allow smaller schools and thus lower threshold populations (Bray 1987). In low density rural areas of Europe and North America in the nineteenth century, school provision was achieved by having much lower thresholds in one-teacher or two-teacher schools. Even today, despite major improvements in transport and individual mobility, many rural schools in these countries continue to have small enrolments, enabling small communities to retain a school with a small number of pupils but normally satisfactory pupil/ teacher ratios, with each teacher teaching two or more age groups simultaneously. Multigrading has been, but usually by default, the classic solution to the small school problem throughout most of the

Third World, but it is not popular with teachers, for they need to be better trained in the very specific methods of handling children of different ages and levels. Furthermore, teachers do not normally prefer to work in small, isolated communities. Multigrading is felt by teachers and by communities in general, unnecessarily in view of much developed world experience, to be rather a second best, providing a less than adequate education in comparison to the full grade school. Other strategies for lowering the threshold populations include having biennial intakes (i.e. having new cohorts in the school every second year instead of every year, and having, for example, three classes and three teachers at any one time but enabling access to a full six-year cycle), or having collaborative sharing of teachers and other facilities in relatively nearby schools, the so-called *nucleo* system in Latin America (Hallak *et al.* 1976).

In practice it has proved as difficult to restructure schools to lower the threshold as it has been to extend the range. Children in large areas of many countries remain beyond direct daily access to the education system. For a few countries the preferred strategy has been settlement nucleation – to bring pupils to the schools rather than the schools to the pupils. In Tanzania, for example, there were two basic reasons for the villagization programme of the mid-1970s. One was intrinsically political, to promote *ujamaa*, the socialization of rural production. The other was to provide better access to public services, notably health care and education (Thomas 1985). The threshold population for a primary school in Tanzania is 1300 (Gould 1982a), and the mean size of the *ujamaa* villages is over 1500, so that a full primary school is appropriate for most villages. Most Tanzanian children now have access to at least the primary school cycle. However, enrolment rates do not reach 100 per cent at any level, in part because there have been insufficient teachers and insufficient funds to employ them and in part because of lack of demand for schooling among what Hyden (1980) has called an 'uncaptured peasantry'. Improving physical access is important for equity objectives, but is not in itself sufficient to ensure enrolment or to reduce drop-out.

Technical and political conflicts

These planning approaches to school provision are based on positivist assumptions. Planning is taken to be a purely uncontroversial technical matter, operationalizing a universal consensus over needs and priorities. Since schooling is a vitally important issue at all scales and therefore a matter of concern to all sections of the population, political authorities and decision makers may prefer priorities that are not the same as those suggested by technical efficiency criteria as seen from within the education system. Furthermore, consumers are all assumed to behave 'rationally', and in particular always to attend the school nearest to their home in order to minimize travel costs, both direct costs and forgone opportunity costs, by minimizing the time taken on the home/school journey. In this, therefore, consumer behaviour is governed only by distance, and is indifferent to the quality of provision at any of the outlets or facilities. This may be a valid assumption when planning water supplies (it may be assumed that the water at standpipe X is as good and pure as the water at standpipe Y so that choice of standpipe is governed only by distance), but may be quite inappropriate for schools where the actual and perceived quality of the education provided can and does vary substantially from school to school. Thus for both reasons – of political concerns and of consumer choice – the geography of local school provision is much more complex than is assumed by the rather simplistic facilities planning approach. The conflicts between a delivery system based on a purely technical approach and one which takes political, community and individual choices into account are resolved in the working out of the equity/efficiency dilemma.

The local political context, in which decisions about locations of schools and allocations of inputs into the system are made, can vary very substantially. Where there is a vibrant local democracy and strong institutions of local government which have a local responsibility for resource allocations then the political process will impose additional criteria on the technical aspects of educational planning. Criteria of need and priority may be established on a political basis rather than on an objective assessment of absolute deprivation. While it may be technically desirable to direct new or additional resources to an unserved or underserved community to reduce intra-district inequalities, political pressures may dictate a different set of priorities, still within national guidelines, that will direct resources to communities that still have an absolute shortage of school places. The local political process often exacerbates local inequalities: every politician wants a school in his or her constituency, and politicians with most influence are often able to prioritize resource allocations in a pattern that might not satisfy the priorities derived from technical criteria. Even where there is no overt local political process in the form of local government, the technical officers of a Ministry of Education responsible for local allocations are influenced by personal contacts and loyalties. They may be simply browbeaten by a forceful headteacher into allocating new teachers to a school or by a strong community lobby into allocating a new school to that community, or upgrading a primary school to junior secondary status.

Objectively defined needs and priorities in these circumstances are modified, formally or informally, by the political process. The extent to which modification is possible or common, varies widely in the Third World. In some countries the political criteria are all-important and there is only a weak presumption of technical objectivity at the local level. Allocations may be largely a matter of crude political clout, as was noted above for Nigeria (Chapter 3). More commonly, however, and with increasing importance in countries with a national ethos of equality of opportunity that extends beyond rhetoric into the everyday activities of government, there is a strengthening of thc technical presumption of objective allocation criteria of needs assessment and demand. It is supported by an increasingly large and better qualified professional staff to generate allocation proposals based on these criteria. Even these proposals may remain open to political manipulation, but where such technical proposals are made and known, the scope for inappropriate or self-interested allocations is much reduced.

Local communities themselves may also have priorities that are different from those incorporated in the national technical systems. Universal primary school provision may seem a lesser priority at the local level than ensuring more local children gain access to secondary school, even where national priorities, and therefore the assumptions built into the technical systems, emphasize the initial access. Indeed local communities may wish to give more or less priority to education compared with other services over which they exercise control, as in making the choice between a school or a clinic, for example. In circumstances where there are sectoral rather than aggregate composite allocations of funds to local districts, with an implication of no discretionary allocations between sectors, a rigid technical basis for allocations will be stronger. Increasingly, however, as local authorities gain more financial and administrative discretion from central control, the different needs and priorities of local communities can be recognized. As local population and economic circumstances vary, so too will the priorities for allocations to individual school and communities vary.

The assumptions of technical 'rationality' are most seriously undermined at the local scale by the issue of individual choice. Distance-minimizing criteria assume that all schools offer the same service. They do in formal terms, in that they offer the full curriculum for any given level, but the quality of that education can vary widely from school to school. Quality, in this sense, can be measured in terms of inputs (e.g. qualifications of teachers, number of textbooks, quality of buildings) or, more controversially, in terms of measured outputs such as performance in national examinations. In many countries the proportion of children passing an examination at the end of a cycle or the admission examination to the next cycle is a much discussed index of 'quality'. Children are more likely to be able to or prefer to travel further to a better school than to a less good school, and so catchment areas can be differentiated by quality where children have choice. The range of higher quality schools is in effect extended, though in the case of primary schools to which most children are obliged to walk, only by a small amount.

In most rural areas of the Third World there is not in practice a choice, as attendance at the nearest school may be the only feasible possibility. In areas of higher population densities the choice may be a real one with two or even more schools within the range. Overlapping catchment areas are common in such areas. The catchment areas are differentiated not only by distance but by perceived quality with the 'better' schools having larger catchment areas than the 'less good'. They will be better able to exercise a stronger control over admissions, either by charging higher fees or by having some other admissions criterion. The positivist planning model that is the basis for school location planning assumes, unrealistically in such circumstances, that distance is the only differentiating criterion and that catchments are effectively neighbourhood catchments. Since school quality does vary so much within a small area, and especially within towns, the catchments are very strongly socially rather than spatially differentiated.

The most familiar form of social segregation of school catchment areas is by sex, with separate boys' and girls' schools. This is particularly characteristic of Islamic societies, though by no means all Muslim countries have separate boys' and girls' schools, even at primary school level. Traditionally, boys had better access to Koranic schools, but boys and girls could be together for basic religious education. For more formal secular schooling many Middle Eastern countries have seen the development of separate systems for boys and girls, even at primary school level, often with two schools side by side, with two sets of teachers and equipment as well as pupils, and differentiated only, as in Pakistan, by a perimeter wall round the girls' school. In some Islamic societies, however, notably in Malaysia and Indonesia such differentiation has not been common, suggesting that the basis for the differentiation is cultural rather than strictly religious in so far as these can be separated. A tradition of separate boys' and girls' schools has also been associated in Africa with Christian mission schools, especially in boarding schools at secondary level.

While there may be a cultural and even technical justification for separate boarding schools, the existence of separate boys' and girls' systems can introduce a major problem of high per capita costs for primary schools in rural areas. In the Middle East, for example, rural population densities tend to be low overall, but to be concentrated in nucleated settlements around water points. It is not unusual to find two relatively small schools, the boys' rather larger than the girls', in such villages, but both schools operating at well below enrolment efficiency, with low pupil/teacher ratios, especially in girls' schools. Girls' schools in such rural environments have an additional problem to attract qualified female teachers in societies where the independent mobility of married and unmarried women is severely constrained. Separate boys' and girls' schools therefore mean not only that costs are raised, but that there are further constraints on ensuring girls have access to good quality schooling. These constraints exacerbate the under-achievement and general low status of women in these circumstances.

In such an area there is relatively little 'choice' in a cultural sense for the individual. The choice is one dictated by society as a whole. In other cases, however, parents and children may have a more obvious choice. This will be where schools are differentiated by language of instruction (e.g. where several indigenous languages are used in lower primary schools in ethnically mixed communities, as in the major towns) or, much more commonly, where schools are differentiated by management authority – public or private – or even within the public sector by denominational affiliation. In the public sector in Ghana, for example, there are government schools, wholly funded and managed by district authorities, and also schools managed but only partly funded by over ten religious groups, Christian and Muslim. These schools cater mainly for the adherents of these groups. They are to be found not only in urban areas, but are familiar in rural Ghana, such that there is a complex pattern of overlapping catchments. This has been shown to be familiar in many settings, not only in Africa, e.g. in Uganda (Gould 1973), but elsewhere, as in the complex religions and denominational social environment of Lebanon (Khoury *et al.* 1975).

'Quality' in such situations is defined in terms of the religious component of the 'education', not normally formally in the curriculum, but certainly in the 'hidden agenda' of each school, and the basis on which schools are differentiated in the public mind. From a technical point of view, the ability and willingness of parents to exercise the choice to send their children to a school that is not necessarily the nearest school may create problems of enrolment projection for individual schools, even though aggregate demand can be calculated from census and other standard sources. More serious problems arise because the denominational rivalry tends to generate many small schools that operate at sub-optimal pupil/teacher ratios. Substantial savings that can be directed towards qualitative improvement, such as more textbooks and other learning materials, and overall improvements in learning standards can accrue from school amalgamation. These can be achieved only at a perceived cost of loss of ethnic or denominational control and identity. Public systems normally seek to encourage rationalization of small schools with overlapping catchments. However, amalgamations often meet with considerable community resistance, especially where the additional marginal costs of small inefficient schools are not borne directly by the communities themselves, but by the public purse, as in Ghana, where the government employs and pays the teachers in such schools. Reducing costs and raising objective quality in these circumstances will depend on the political will of government to tackle the problem of sharp inter-school per pupil cost differentials by ensuring pupil/teacher ratios rise throughout to nearer the national norm. This will require amalgamation of some neighbouring and formerly rival schools.

The largest local inequalities are usually to be found between the public and private systems, where choice has allowed catchment areas to be differentiated by socio-economic criteria, notably ability to pay. In urban areas in particular, the schools system often mirrors the urban social environment at large: sharp socio-economic differentials which are not always spatially differentiated, with private schools catering for the rich who prefer not to send their children to seriously under-funded public schools. Here again the schools system is mirroring some of the broader issues in the urban social fabric. National goals of equity through education and equality of opportunity for all children are seriously compromised by differentiated access based on social rather than spatial criteria. Positivist assumptions of schools planning are not wholly operational unless there is broader political commitment to promoting social equity.

Schools and the rural community

The local geography of education must consider not only the location and catchment areas of the schools, but the impact that these facilities have on the life and work of the communities in which they are located. In towns the direct link between school and community is obscured by the existence of many other facilities and enterprises that have their overall effect on the urban area, such that a separate school/community link is not immediately apparent. By contrast, in rural areas schools have an immediate and recognizable impact. The most immediate impact is a visual and physical one: the school is usually centrally located within the community, occupying a fairly large site, and often architecturally distinctive. In some countries primary schools are built of traditional materials in traditional building styles, but increasingly, and certainly for secondary schools and above, the schools tend to be built to standard designs of the Ministry of Education using at least some manufactured rather than traditional materials. In very poor communities this may mean that physical standards of school buildings are much higher than in most other buildings locally, with an expectation of services, such as mains electricity and running water where schools have laboratory subjects, and standard requirements and designs for these buildings. Immediately there is implicit in this a tension between the school as a traditional feature of the community, conforming to its norms and values, and the school as an innovation, bringing new norms and values to which the community might aspire.

Schools have a social and political impact. It is the school as an innovation, the most obvious artefact of modernization, that most commonly attracts the interest of rural communities. It is often the only tangible link the rural community has with the state and the services provided by the state. It therefore represents a world outside the community that links it with the broader national or regional economic and social aspirations for development that are wrapped up in the ideology and rhetoric of the state. It is a manifestation of progress and of the benefits the state brings. Even where the physical costs of construction have been borne by the community and the building is in a traditional style, the school is still seen to be a symbol of and a mechanism for a new, and in general terms, better lifestyle. This prestige is of course of great importance to the political process, and what goes on in the schools therefore becomes an issue of great community concern, even amongst those who do not have children in the school. The teachers and particularly the headteachers are normally accorded prestige and authority as a result of their status, inevitably becoming leaders of community opinion, especially in matters of modern development and change. Rural teachers are often among the wealthiest members of the community. They generally have experience of living, working or learning elsewhere, and especially in towns, and tend to assume the status of role models for many parents as well as children.

Although rural schools are accorded great prestige by the communities themselves, in practice they tend to reflect broader changes and tensions in society as a whole. The classic discussion of the relationships between a school and a rural comminity is Roger Thabault's *Mon village*, a study of 'education and change in a village community', Mazières-en-Gâtine in Vendée in France, 1848–1914 (Thabault 1971). Thabault was Director of Schools in French colonial Morocco in the 1920s and the study of his home village and its school over that period of very great social and economic change in France, he argued, shed light on the particular problems of the effects of schools in Morocco and, by implication, in other similar countries. He showed how in the earlier years of his study period the village school, imposed by government and local notables, had little impact on the lives and aspirations of the local population. The impact became substantial only when the village economy became more closely integrated into regional and national markets, and in particular after the coming of the railway. Then parents and children saw the school to be necessary for taking advantage of the new opportunities in the wider economy and society:

> The Mazières school has thus seemed to me to have been essentially urban and universal in outlook, not rural and localized. At every turn, however, it seems to have depended very closely on the environment in which it existed. The intentions of central government became practical reality only in so far as the people's minds were prepared for them and when they assumed a form acceptable to them. The dilligence of pupils and the influence the school exerted on them were at all times dependent on the economic and social state of the commune and the attitude of mind of the people. It did not create the belief that it helped to spread; but it was the mirror which reflected and focused this belief.
>
> (Thabault 1971:232)

The previous section of this chapter has described how catchment areas of individual schools may be differentiated by social characteristics. Differentiation by religion or denomination, related to the founding or managing body of the school, is common in multi-faith rural communities, but of more long-term and more potentially socially disruptive significance is the differentiation of school catchments by socio-economic criteria. Better schools, often but not always private schools, provide a higher cost education to the better-off, with objectively higher standards in terms of examination results of pupils compared to the children of the poor who attend less well-endowed schools. The existence and strength of private schools and their socio-economic selectivity and resulting social impact is of course part of a broader social ethos concerning the nature of inequality in the society in which they are found. As noted in Chapter 3, in socialist Tanzania the government tried very vigorously, but ultimately without success, to prevent the growth of private schools, but demand for them, especially in the richer areas of the country, notably Kilimanjaro Region, was such that they have grown with potentially serious effects for the national strategy of promoting equity through the education system (Samoff 1987). In capitalist Kenya, by contrast, not only is there the large system of *harambee* community self-help schools, but at the secondary level a large number of government as well as private secondary and primary schools which have high fees and cater for many of the élite, and are much encouraged by the national ethos, if not by direct subsidy from the education budget of the Kenyan state (Court and Kinyanjui 1980).

In rural communities schools may have a significance that takes them into the controversies of local political rivalries. It is not unknown (e.g. in Pakistan) for schools (and their teachers) to be closely identified with a particular political party, and for catchments to be differentiated by political allegiance. However, even in formal one-party states, local rivalries within that party may be reflected in appointments of teachers to schools or promotions to headteacher, or in appointments of school governors or managers. Such appointments can bring prestige as well as access to some resources and institutions of the state. The school becomes an arena in which these broader rivalries and tensions can be played out. They are all the more important where school managers have significant local responsibilities (e.g. in hiring and firing teachers).

In their economic impact on rural communities schools can be of major short-term and longer term significance. Chapter 3 considered how schools can be viewed as economic phenomena in their own right, creating employment for teachers and other staff, stimulating the local building industry in school construction and for school furniture, stimulating local services either directly (e.g. clothing and dressmaking activities where school uniforms are required) or indirectly through the multiplier effects of expenditures by staff. The education 'industry' can be a major generator of economic activity and income. More commonly, however, schools are taken to be part of the infrastructure, items of 'social overhead capital', rather than directly productive activities. Their economic impact is assumed to be indirect and enabling rather than direct or immediate. In some cases schools may be formally involved in production. There may be school farms or gardens or other activities producing food to feed the pupils or else directly for sale. 'Education with production' has been a familiar phrase in the schools in some countries, e.g. in Zambia (Achola and Kaluba 1989), in the 'brigades' in Botswana (Van Rensberg 1984) and in Cuba (Figueroa *et al.* 1974; Richmond 1985). In other cases the school year can be made to be consistent with the local production needs to ensure availability of children to work in the fields at peak labour periods, notably for weeding and at harvest, by closing the schools at critical periods, either through informal local arrangements of by national or regional decisions about school holiday periods. In general in rural areas, school holidays have been organized to be consistent with the demands of the agricultural calendar.

Of much more fundamental importance to rural communities is the indirect effect of the education they provide on the economic life of the community, and here again the effects can be both positive and negative. Most positive is the fact that there seems to be a relationship between education and farmers' income: in any given rural community the more educated have higher incomes from farming. World Bank research in particular has provided the empirical justification for investments in education as a support for rural development: 'Four years of schooling is capable of enhancing the output of modernizing farmers by as much as 10% per year, compared with uneducated farmers in the same area' (Colclough 1982:177). Even at low levels of educational attainment (i.e. up to only four years in school) there are large returns to education amongst small farmers (Fig. 4.2). Returns to basic education are stronger in more modernizing environments since there is a strong commercial incentive to better yields. In such a context the effect of basic schooling may be not merely to allow the benefits of literacy and numeracy to apply, but also to expose pupils more readily to a wider world of knowledge and experience that will

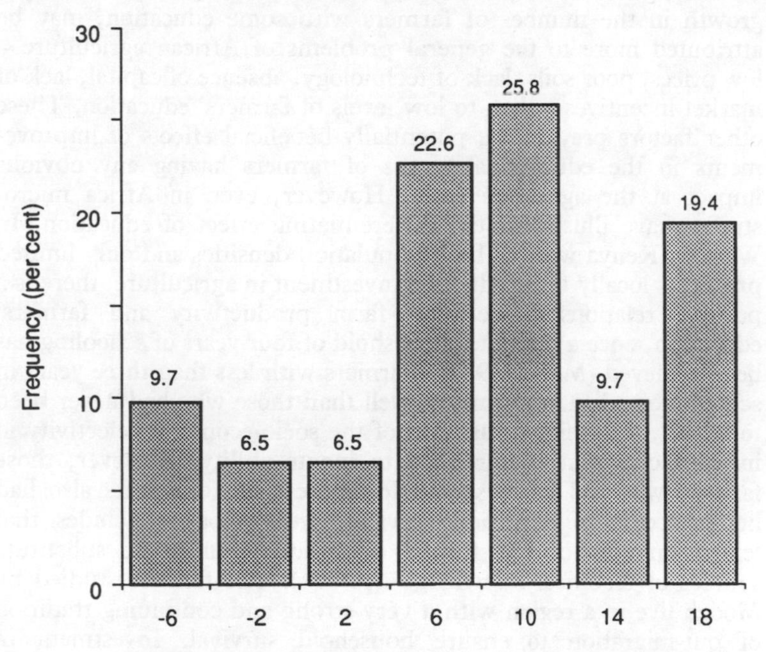

Note : Mean, 8.7 per cent; standard deviation, 9.0 per cent

Fig. 4.2 Results of 31 data sets relating schooling and agricultural productivity (unweighted) (from Jamison and Lau 1982:9)

accelerate any more general tendency for innovation and modernization. Nevertheless, it seems that, other things being equal, a basic education will provide a farmer with a basic literacy (sufficient to read instructions on fertilizer or insecticide containers), and a basic numeracy (sufficient to permit measurement of crop levels or fertilizer inputs). In Nepal farmers in the lowland *terai* region who had been to primary school were more likely than those who had not to adopt chemical fertilizer, and levels of numeracy affected wheat productivity in particular (Jamison and Moock 1984). Higher levels will introduce more sophisticated benefits (Table 4.2). In this area, as elsewhere in Asia (e.g. Thailand), there are close interrelationships between educational achievement, farmers' incomes and nutritional status (Jamison and Lockheed 1987).

Schooling can therefore reinforce other local factors in improving agriculture, but may not begin to have an effect where other factors or incentives are absent. That agricultural production has not generally expanded rapidly in Africa, despite a rapid growth in the number of farmers with some education, may be attributed more to the general problems of African agriculture – low prices, poor soils, lack of technology, absence of capital, lack of market incentives – than to low levels of farmers' education. These other factors prevent the potentially beneficial effects of improvements in the educational status of farmers having any obvious impact at the aggregate scale. However, even in Africa micro-studies have illustrated the differentiating effect of education. In Western Kenya with its high population densities and only limited prospects locally for productive investment in agriculture, there is a positive relationship between farm productivity and farmers' education, once a minimum threshold of four years of schooling has been achieved (Moock 1981). Farmers with less than three years in school seemed to perform less well than those who had never been to school, this being a function of the socio-economic selectivity in initial access that is unrelated to innate ability. However, those farmers who did go to school for more than four years also had better access to extension services, and Moock concludes that 'extension contact and formal education seem to be substitute sources of technical knowledge' (p. 734). The farmers studied by Moock live in a region with a very strong and continuing tradition of out-migration to ensure household survival. Investment in education in that area is seen to be primarily a means to migration and an urban job (Gould 1985a, 1988a), but it clearly also has an impact in the rural economy.

To raise the question of the migration of school leavers, as in the Kenyan case, is to introduce a much more fundamental criticism of the negative impact of schooling on rural communities:

Table 4.2 Four basic stages of agricultural productivity and their learning requirements

Farmer-entrepreneurs' technology level	Agricultural inputs	Minimum learning requirements
Level A: Traditional farming techniques passed from parent to child	Local varieties of seeds and implements	Addition and subtraction – not necessarily acquired through formal education
Level B: Intermediate technology	Small quantities of fertilizer	Addition, subtraction, division, and rudimentary literacy
Level C: Fully improved technology	High yielding varieties: proven seeds, rate of application of seed, fertilizer, and pest control per acre	Multiplication, long division, and other more complex mathematical procedures; reading and writing abilities, and rudimentary knowledge of chemistry and biology
Level D: Full irrigation-based farming	All above inputs; tubewell access during the off-season and water rates per acre	Mathematics, independent written communication, high reading comprehension, ability to research unfamiliar words and concepts; elementary chemistry, biology, physics; and regular access to information from print and electronic sources

Source: Heyneman 1983.

the view that the main impact of education has been to undermine the local economy and to prevent its improvement. This has happened as a result of two interrelated processes. In the first place the attitudes and values introduced by education are essentially Western and urban values, as Thabault argued, and the skills that formal education provides are skills that have greater reward in the modern urban sector than in the rural sector. Even basic numeracy and literacy as well as more sophisticated skills are more essential in the modern commercial economy. However, it has been argued from a study of rural education in Sri Lanka, where the general level of basic education is high, that rural people do value basic education for its own sake and for its value to them in the rural areas, not least because it makes rural people less vulnerable to being cheated by urban traders:

> Parents are more realistic than is sometimes supposed; while they hope for higher education and good jobs, they are prepared to settle for less . . . success in the economic sphere was not the primary motivation for going to school. Rather it was education for general knowledge and enrichment or education to avoid extreme vulnerability in the modernizing economy that was foremost in their minds.
>
> (Baker 1989:517–18)

Urban skills bias is inevitably related to the second process, that of migration. The educated acquire skills that are rewarded in urban and commercialized rural areas. They are disproportionately more likely to become migrants and to seek an alternative to the rural economy as their principal source of livelihood. Where there is a sharp difference in male and female participation rates, it is likely that the males migrate and the women stay behind, as in central India. *Men to Bombay. Women at home* is a provocative title of a study of a village, Sugao, on the Deccan plateau in Maharashtra (Dandekar 1986). The migration differentials it describes are reinforced by differential attendance and achievement in the village school. The literate proportion of Sagao's population rose from 20 per cent in 1942 to 67 per cent in 1977, and there is great demand in the village for additional schooling. Most of the male migrants, the majority of whom find work in the textile mills of Bombay, see schooling as an essential prerequisite to becoming a migrant. A high-school diploma is essential even for textile workers. For girls the value of education is rather different. Fewer girls than boys complete primary and high-school cycles, but the proportion of girls has been rising even though there are fewer income earning opportunities for them, both in the village and beyond. There are a few successful educated women in urban jobs, but these are very much the exception. It seems that:

> . . . the education of women appears to be a response to an increased demand

for educated brides. A Secondary School Certificate is considered valuable in arranged-marriage negotiations with educated grooms. A college education, on the other hand, appears to be a disqualification, since it is difficult to find a sufficiently educated mate in the rural network.

(Dandekar 1986:98)

It is undoubtedly the case that the main thrust of schooling has been to introduce national or even international concepts of what is 'modern' and appropriate to the pursuit of economic and social development. Schooling inculcates the values of universalist science and objective rationality, apparent in non-agricultural production and in urban areas. It is in these occupations and in these areas that 'success' is to be sought. The urban élite are the implicit or even explicit role models for school children. They are not encouraged to wish to become subsistence farmers, working hard for a meagre and uncertain return from the land when the obvious alternatives seem to be being made accessible through education. The further they progress in education, the more they acquire skills and knowledge that seems to fit them for the urban commercial economy, and the less they will be satisfied by seeking a future in traditional activities in rural areas. There can be no denying that the educated are disproportionately more likely than those with less education to seek alternative economic opportunities and to migrate outside the rural community (see Chapter 7). The issue that needs to be raised in this chapter on education and the local community is the extent to which these attitudes and associated migrations of the educated have an adverse effect on the rural area.

Vocational schools and rural development

Much of the criticism has been levelled at the curriculum of the schools. It is felt to be too 'academic' or 'non-practical' on the one hand, and at the same time inculcating skills and values appropriate to the modern economy on the other. Colonial education certainly had a strong presumption against traditional values and activities, but even in the colonial period there were many schemes for rural-based and practical education, e.g. in the 1920s in the British colonies in Africa (King 1971; Sinclair with Lillis 1980). In more recent years many countries have sought to introduce curriculum change as a means of developing more positive attitudes to the rural economy, but attempts to 'ruralize' the curriculum and introduce practical subjects have been conspicuous by their small impact.

Philip Foster in the mid-1960s, based on his own experience in Uganda and Ghana and on the general experience of colonial Africa and Asia, formulated the idea of the 'vocational school fallacy', that powerfully exploded the curriculum bias argument (Foster 1966). Rural-based vocational education, he argued, was in fact a common objective in Africa and Asia, but the population regarded it as a second best, designed to 'keep the natives in their place'. They wanted to have the opportunity to be involved in the modern economy, to acquire a highly paid skill and to work in town. Attempts to ruralize the curriculum, however well intentioned, flew in the face of economic reality and perceived economic opportunities in the earlier period, as they continue to do at the end of the twentieth century. Many Third World countries have sought to introduce a more 'relevant' curriculum with agriculture and rural science as formal subjects, but everywhere non-rural subjects are still preferred where pupils have a choice. Foster argued that academic subjects are in practice 'vocational' since the preferred jobs were in white-collar urban occupations. Literacy and numeracy rather than specific technical skills are the basic requirements for a civil servant or other white-collar worker. Apparent anti-rural bias in the curriculum is a reflection of broader structures of prestige and reward in society, and not to be resolved through changing the curriculum without changes in these basic structures.

The strength of the 'vocational school fallacy' is amply demonstrated in a wide range of experience, but nowhere more obviously than in the youth polytechnics in Kenya. By 1972 53 youth polytechnics (at that time styled 'village' polytechnics) had been established. This number had risen to 195 in 1978, to 321 by 1985, and 545 in 1989 (Kenya Government 1989b:221). There are over 30 000 trainees in these polytechnics, mostly primary school leavers and secondary school leavers and drop-outs. The demand for rural training in independent Kenya can be traced back to the general malaise felt at independence in the early 1960s about the nature of education for rural development in Kenya (Sheffield 1967). The early initiatives to promote rural education were spontaneously taken up in several types of self-help *harambee* efforts by churches, communities and cooperatives, but not directly by any government initiative, though most polytechnics now receive some (variable) government financial support. The polytechnics as a result vary greatly in their size, sources and amount of financial support, in their buildings and other facilities, and in links with the local community. They are generally considered to be non-formal institutions, though the element of formal education, with regular terms and timetable and preparing students for formal examinations, is a very evident and common feature (Ayot 1987).

As far as the Government of Kenya is concerned youth polytechnics have several major objectives, amongst which are the need to train school leavers for work in their home areas and, obviously linked, to develop skills and attitudes amongst boys and girls that will lead them to income-earning activities in rural areas. The obviously instrumentalist view of the function of the polytechnics is responsible for their being given prominence in the general literature on education for employment. David Court, however, believes they are of more fundamental importance:

> . . . the long-term significance of the polytechnic movement lies not simply in its reflection of *harambee* or its role in vocational training, but in its potential as an educational experiment. Because village polytechnics exemplify a diversity of activities, techniques and organisations, it is more appropriate to treat them as an ideological movement than as an institutional prescription. . . . Thus the essence of the ideology can be seen as an attempt to break away from conventional concepts of academic schooling and to develop types of training that are rooted in practical need and which convey a sense of individual purpose and capacity for continuing self-instruction.
>
> (Court 1974:220–1)

The public and individual perceptions of this type of non-formal skill training are premissed on the assumption, consistent with the wider context of Kenyan development ideology, of individual opportunity differentiated by individual abilities, both inherited and acquired. That a strong shadow system has emerged confirms a dissatisfaction with the formal post-primary education sector, yet the demand for formal secondary education remains strong. Polytechnics are felt to provide a training that is second best, for formal sector opportunities for which academic qualifications are required remain the most attractive option for young people.

Courses offered in village polytechnics indicate the range and emphasis in skill training that is provided. The majority of courses are in carpentry, in masonry, in tailoring and dressmaking and in home economics. Explicitly agricultural courses are offered but they are not common. Each of these subjects clearly embraces skills required in rural areas at the low level to which the skills can be taught in a two-year course to the level of the Grade III Trade-Test Certificate, except in agriculture for which there is no formal qualification. There can be little doubt of the need to raise the skill levels of carpenters, masons, tailors and farmers in rural Kenya, and the village polytechnics seek to make a major contribution in this direction.

However, two major factors intervene to considerably compromise the potential success of such a contribution. Within rural areas themselves the levels of demand are such that the services of skilled people cannot be afforded. Rural poverty is so widespread in Kenya that there is little effective demand for many services, and this is particularly so in the poorer areas of the country. There is a substantial off-farm rural employment sector in Kenya as a whole, encompassing over 1 million people, as many as are employed in the formal sector, but the pattern is sharply differentiated geographically. There is a wide gap between Central Province, the area of the most vibrant rural economy, and, for example, Western Province, where annual average value of non-farm activities per household in 1974 was only one-third the national average and less than 20 per cent of the Central Province average (Freeman and Norcliffe 1984:229). In less prosperous areas the capacity of the local economy adequately to support the services offered by a regular supply of polytechnic trainees is likely to be small.

The other major factor is the continuing attraction of urban areas. Not only are the skills the trainees acquire also sought after in urban areas, but the level of demand for such services and the level of wages are very much greater than in most rural areas. It is not surprising therefore that most studies of polytechnic leavers, both in the earlier period (Court 1974) and more recently (Barker and Ferguson 1983; Gould 1989), have shown that towns and particularly Nairobi are a major attraction. The trainees may not necessarily or even normally use the skills they have acquired when they do eventually find a job in town. The period in polytechnic training is, for many of them, a means to an urban end, though there is, as yet, no evidence to suggest that polytechnic trainees are any more or less likely to find an urban job than their peers who did not continue in any form of training after primary school. Youth polytechnics can, in theory, provide skill training that can be of considerable benefit for rural development and may be in many cases. However, the trainees tend to use the skill training to open further their own spatial horizons by migrating to urban areas where they are more likely to be able to use these skills with profit to themselves, especially in the informal sector.

Nevertheless the search for a 'relevant' rural education continues in Kenya, both in formal and non-formal institutions. This search has been a recurring theme in reports of international agencies in Kenya as elsewhere. A report entitled *Rural development, employment and incomes in Kenya*, prepared as part of ILO's Jobs and Skills Programme for Africa (JASPA) in the early 1980s recommended that: 'attention should be directed towards agriculturally-orientated or rural-orientated school activity and curriculum content as a means of stemming rural–urban migration as well as increasing agricultural productivity in the long run' (Livingstone 1981:20). Village polytechnics have sought to meet these objectives, but have clearly not achieved them. As an experiment in rural education they have achieved similar disappointing results to many other experiments in rural education elsewhere in the Third World.

In addition, however, it has to be recognized that the overall impact of schools, regardless of curriculum, is only part of a broader range of causes in rural out-migration. The area rather than the schools is the more important. One study of primary school leavers from ten schools in two sub-locations in Western Kenya showed that there was some variation by school of migration attitudes and behaviour (leavers from the better and more prestigious schools were more likely to become migrants). School attended also had a great effect on examination results. However, inter-*school* variation was less than the inter-*area* variation when all the pupils in each of the two sub-locations were compared. Area rather than school affects migration behaviour; school rather than area affects formal achievement (Gould 1983).

Local management issues

It is therefore clear that at the local scale the fears that schools constitute a threat to the rural economies of Third World countries is somewhat misplaced. Schools can contribute very positively to rural production, but that contribution can only be seen in the overall context of the political economy of the state and the broader questions of rural–urban and regional balance. Schools can facilitate change and promote production within an enabling political and economic environment; indeed, they are part of the infrastructure of that environment. They are not, however, sufficiently strong or critical to have an effect that is inconsistent with that broader context. The fact that most people in most rural communities welcome schools and strive to send their children to school for as long as they can afford to, confirms that the school is seen by rural inhabitants to be a positive rather than a negative force in rural development.

Since schools seem to be much in demand for both social and economic reasons, they are important to rural communities and therefore their management becomes an issue of substantial local significance. Normally the schools are seen as the responsibility of the state, provided from outside the community and managed in a top–down hierarchical structure from central or regional government. There is often little or no local involvement. If there is any community involvement at all it is in fund raising or in donating time and materials to build classrooms or other facilities in the school. Local communities normally have no direct influence over appointments of teachers, choice of curriculum and income or expenditure patterns of the school. As education systems have expanded rapidly in the past few decades it has become increasingly difficult to maintain a centralized, top–down management structure. Most education systems have experienced some deconcentration of management to lower tiers of authority. There is less rigid central control over income and expenditure, especially in those countries affected by policies of 'cost sharing' in education and where an increasing proportion of revenues needs to be generated locally. In these cases the demand for much more of a bottom–up system of management could be more realistic, with local control of aspects of curriculum (e.g. what would be the language of instruction at the initial level, which 'practical' subjects to teach at higher levels), or when to have the school vacations (to fit with the local agricultural requirements for children's labour).

Thus a conclusion of this consideration of the education system at the local scale, and particularly of the place of the school in rural communities, links with the concluding discussion of the previous chapter in its recognition of the advantages of local control in a decentralized system. Where there is some element of decentralization with some local inputs into education decision making, then people in these communities – parents, children and even those with no direct link with the school – will be more likely to be involved in defining and implementing local needs, and to see the school as contributing to achieving them. The school will then be more likely to be an active force for local economic and social development.

Chapter 5

Education and population growth

Formal education through schooling is directed primarily at children. Since Third World populations have a much higher proportion of young people than is the case in more developed countries, the educational needs are proportionately much greater and expenditure of governments on education needs to be proportionately much larger. At its extreme, more than 50 per cent of the population of some African countries is aged 15 years or less, though the range is more typically nearer 40 per cent. In Europe that proportion is less than 25 per cent, and was less than 30 per cent even at the period of maximum growth in the mid-nineteenth century. These proportions are a direct function of population growth. Where rates of population growth are high – typically between 1 and 1.5 per cent p.a. in Latin America, the Middle East and Asia, and in excess of 2.5 per cent in Africa – then the contrasts in age structure are exaggerated. Third World countries typically have a large proportion of young people but a small proportion of old people as their 'dependent' population, i.e. as a proportion in the main economically active age groups of adulthood.

The implications of a 'dependent' population, familiar in population analysis, refer particularly to the resources required by and provided for the young and the old being created by the activities of the non-dependent population. In an industrialized society 'dependents' and 'non-dependents' are much more easily differentiated than they are in Third World countries. In the Third World children and the old have economic and domestic roles that are not incorporated in a Western notion of dependency. Nevertheless, the young and the old do increasingly consume services that require formal government expenditure. The largest and most important of these services, as discussed in previous chapters, is education.

The lives of children and the attitudes of their parents towards them are fundamentally affected when they attend school. Without school to attend, children are more able to contribute to the activities of the household – in domestic work, notably fetching water and looking after younger children; and in economic activities, particularly in agriculture, where children can have a critical role from an early age in weeding or scaring birds or other pests from fields, or in looking after animals. Where several hours of perhaps 200 days per year are spent in school, children's potential role in the household is less, though perhaps as much may be expected of them in the now reduced time available. Not only can they do less, but in going to school they may incur costs of various sorts, directly for tuition and books and indirectly for uniforms, etc., as well as in opportunity costs relative to the other activities. Children therefore become more of a 'burden' on the household economy, more 'dependent' on income and activities of the adult members of the household. In this sense education increases the applicability of the concept of demographic dependency as applied to children. They are perceived as a 'burden' on the state as well as on the individual household. This economic 'burden' has been willingly borne in most countries by parents and by society at large, as has been discussed in earlier chapters. It has been seen as a short-term necessity for long-term benefit, but the calculus of how great that burden might be alters with increased education: for more children in total and for longer for each child who begins in school.

Education affects the demand for children and through that the fertility decisions of marriage partners. One of the main global and most widely discussed effects of education is its contribution to changes in levels of fertility, notably that education seems to be associated with declining fertility – for societies as a whole and for individual parents. The greater the level of education a parent, and particularly a mother, has achieved, the more likely it is that in any given social context the number of children of the couple will be less than the number of children of a less-educated couple. Furthermore, levels of mortality, the other component of natural population change, are also affected by education. Globally, life expectancies of the educated are higher than those of the population who have not been to school, and the children born to educated parents have lower rates of mortality than do the children of uneducated parents. In the past the dramatic falls in mortality rates since the nineteenth century in Europe and North America have been associated with, but not necessarily caused by, rising rates of educational enrolment. So too throughout the Third World there have been rapid falls in mortality rates, and particularly infant mortality rates, in the last half century, at a time when school enrolments were expanding rapidly.

This, the first of three chapters on the effects of educational expansion, examines the relationship between education and population size and growth. It considers the effects of education on fertility and mortality in separate sections, and will bring these together in a final section that considers the possible contribution of recent and future expansions of education to the prospects for future population trends in the various sub-regions of the Third World. It will consider the role of education in accelerating the demographic transition from a high equilibrium of births and deaths, characteristic of traditional societies, to a low equilibrium of births and deaths, characteristic of developed countries. To what extent and for what reasons might education influence the nature and timing of the demographic transition in the Third World?

Education and fertility

How education changes the 'value' of children

In the past in countries of the Third World fertility levels have normally been high. Family size has often been in excess of five children. Total fertility rates (defined as the mean number of children born to women calculated from current age-specific fertility rates) currently vary from 2.4 in China to 8.0 in Rwanda, with a Third World weighted average of 3.9 in 1989, a level that is considerably less than it was in 1960. The falls have been greatest in Asia and Latin America and least in Africa, and in many African countries they have been negligible. Nevertheless, in by far the majority of Third World countries fertility rates are much less than they were in past years. The reasons for this decline are complex and subject to much debate, but they are the reflection of decisions of couples to have fewer children: in the language of economists, there has been a fall in demand for children. In classical economics, falls in demand are to be explained by increases in the *cost* of the item or a loss in its *value* or usefulness so that alternatives become more widely used. Both these apply to the falls in the demand for children, and both are related to the increasing levels of education.

As a result of education the cost of raising children rises. In traditional societies with no formal education, children are 'hands to work' as well as 'mouths to feed', and their net marginal cost is normally positive: they produce more than they consume. They contribute significantly to domestic work in cleaning and food preparation, in fetching water or fuelwood, in looking after younger children. These activities allow adults, their mother usually, more time for the many more specialist and more skilful tasks. In the economic sphere in rural areas children are able to contribute from an early age: scaring birds or other pests from ripening crops; weeding, planting or harvesting; and generally fetching and carrying for adult workers. Family labour has traditionally been fundamental to the rural economy, and children are expected to be available to make their contribution each day. Large families were beneficial; high fertility was a rational response. In Bihar, India, for example, it has been calculated that the benefits begin to outweigh the costs at about nine years, being slightly earlier for boys, and that 'by the age of 16 or 17 the child has already "repaid" his or her parents for the costs incurred up to the age of ten' (Corbridge and Watson 1985:289).

However, education ties children to the school for 4 or 5 hours on 5 or 6 days per week and on perhaps 40 weeks per year. The activities they were normally expected to do must still be done, but either they are done by the children at non-school times or they are done by someone else. The former remains important. Children may have been up and busy for three or four hours before they get to school at 9 o'clock, and after school there may still be a lot to do before darkness falls. Where children are going to school their normal tasks may have to be reallocated, perhaps to parents, more usually to other brothers and sisters – older or younger – who are not in school, but increasingly in a commercialized economy also to paid labour. However the tasks are reallocated, going to school must involve a calculation of opportunity costs. Do the benefits of schooling outweigh the foregone opportunity costs that need to be reallocated?

More immediately apparent than these opportunity costs are the direct costs of schooling that will be incurred. These will include tuition fees and a range of other levies – building fund contributions, textbooks and materials, etc. Education, even if nominally 'free' (and it is no longer even nominally 'free' in a large and increasing number of countries) requires several payments directly to the school or, for uniforms or other materials and equipment, directly by the student. Even if the sums required are small for the initial years, they can quickly grow to very large sums for secondary schools. In Kenya, for example, typical fees in *harambee* secondary schools are K4000 per year ($200), a very substantial sum in a country where per capita GNP is $390. Many households can spend one-quarter of their cash income on school fees. As the direct costs rise, so demand for education (if not for the children themselves) may be expected to fall.

At the same time the usefulness of the children may alter and non-labour substitutes may be found. One of the first impacts of education, as far as parents' perceptions are concerned, is that schooling seems to lessen children's willingness to be involved in family agricultural effort: farming is 'beneath them'. This may be part of the alleged anti-rural bias in the curriculum discussed in

Chapter 4. Parents may in practice encourage some of this by preparing their children, at least implicitly, for a better lifestyle following education. Whatever time is available after school activities (including homework) may not be used, by preference, for traditional family-related activities. This will add further to the overall costs of education, and add further incentive to enhancing production by better use of technology as well as other labour. This may be made possible where, for example, rural water supplies are improved or there is a rural electrification programme, and new technologies can be applied in the home and on the farm to substitute for children's labour.

The high value placed on children – large economic and cultural benefits at relatively small cost – thus alters. The potential benefits are very large, hence the massive demand for education, but these benefits are long-term and may only be realized in 20 or 30 years. However, the initial costs are also large, and need to be paid immediately. The logic of high fertility may appear to be less persuasive. Family size may then exercise a substantial negative effect in the probability that a child will attend secondary school. In Thailand, for example, it has been shown that in conditions of rapid fertility decline more children from poorer households now go to secondary school (Knodel and Wongsith 1991). Smaller families have contributed to a rise in educational attainment amongst the poor, with stronger effects in urban than in rural areas. In urban areas the costs of schooling are more directly borne by parents and the opportunity costs of keeping a child in school are higher. Better to concentrate scarce resources on a smaller number of children – but to ensure that the education that can be supplied greatly improves their life chances – than to spread expenditures thinly among a large family. In Africa, on the other hand, where there is a much wider sharing of the costs of children in extended families and through traditional child fosterage systems, these benefits of lower fertilty for increasing school attendance are much less directly felt (Caldwell 1987). There is an economic logic in the systematic and mutually reinforcing link between education and fertility reduction, which is at the base of a new calculation of costs and benefits with widespread availability of education and a demand for the skills and opportunities it generates. It applies irrespective of what is happening to mortality and increasing likelihood of survival of infants.

Education in population policies

Many Third World governments now have direct and active population policies with a specific thrust towards family planning. For governments the costs of providing education are high and the immediate financial benefits of fertility reduction are obvious. More generalized benefits of reducing high rates of population growth, such as improving food availability or reducing the need to create more jobs, have medium- to long-term impacts, but the effects on the education system are directly felt after five or six years and will accelerate for subsequent cohorts (Chau 1972; McNicholl 1984). Uitto (1989) records how over 30 per cent of the Kenyan government budget was being allocated to the education sector in the 1980s, and, in the absence of any fertility decline, would have been expected to have risen further as the large and increasing proportion of the population in the school-age groups was provided with school places. Since subsequent demographic studies, notably the Demographic and Health Survey of 1989, have shown that fertility rates have indeed begun to fall in Kenya, with an overall fall of 13 per cent between 1978 and 1989, more in some regions than in others and in some regions not at all, education expenditure in the 1990s may begin to fall even if per capita expenditure remains constant. This can create the opportunity to raise per capita expenditure within the same overall budget to attack the problem of low quality.

One well-developed fundamental prop of policies to promote fertility reduction is to seek to create a culture that is amenable to acceptance of increased use of contraceptives and longer birth intervals and better child spacing. This normally involves raising the cost of children, and can be promoted by raising the cost of education. This is one of the arguments used by international agencies to justify 'cost sharing' in education, to pass a larger proportion of expenditure directly to the consumers. One attraction for governments of investing in a family planning programme is that it will reduce the externality costs of large families (i.e. the costs that will be necessary for governments in education and other welfare expenditure) (Chomitz and Birdsall 1991). For families, the increased costs of education, where incomes are low and inelastic, may result in demand for smaller families. In Kenya, for example:

> the long-run trend of increasing education costs per pupil . . . has recently been adjusted upward with the revised curriculum. The accelerating costs, significantly shouldered (given the funding of education) by individual parents largely in proportion to the number of children in school, has significantly increased the already high per pupil costs of education, and has greatly raised the 'price' of children . . .
>
> To the extent that 'cost-sharing' is maintained or increased in Kenya, it seems likely that the force of fertility decline will be maintained for some time.
>
> (Kelley and Nobbe 1990:54–5).

The general demographic effect of the financial policies that are now applied in education in the Third World is widely seen to be to further depress fertility. Studies of the value of children in several Third World countries lend support to this policy presumption. Education has had the effect of removing children from traditional activities, and raising the costs of bringing them up (Lindert 1983).

An equally important and probably more widely recognized demographic effect of education is the relationship between parents' education and both actual and preferred family size. Fathers and, even more obviously, mothers with more education have and want smaller families than those parents with less education. The size of the differential in total fertility rate between the uneducated and those with completed primary school or more (7 years or more) has been shown to be about two children in Asia and Africa, but higher for Latin America (3.43) and the Middle East (4.11) (Cochrane and Farid 1989:45).

The relationship is not perfectly linear, however. Several national studies untertaken as part of the World Fertility Survey (WFS) of the 1970s and early 1980s, identified maximum fertility amongst those with some primary education, a higher level than those with no education at all (Fig. 5.1). In Sub-Saharan Africa, results of ten national surveys of the WFS identified a total fertility rate of 6.99 for those with no schooling, but 7.43 for those with 1–3 years' schooling. This fell to 6.47 for those with 4–6 years' schooling and less than 5 for those with more than seven years' schooling. This irregularity may be due to the fact that those women who did not attend school in most societies tend to be the poorest and least well nourished and to do most heavy manual work, in agriculture and in the home. They are therefore most prone to miscarriages. They experience lower fertility, but not necessarily fewer pregnancies, for a larger proportion of pregnancies do not come to term. However, for those women with some education there is a very strong decline in fertility with secondary and higher levels.

Cross-country comparisons of results of the World Fertility Survey case studies confirm the general relationship, but there are many exceptions (Cochrane 1979). In Indonesia, for example, fertility is *positively* related to education, since, as a result of vigorous and successful family planning campaigns, many small farmers have high contraceptive usage and low fertility but they also have low levels of educational attainment. In general the studies confirmed earlier conclusions about the differential effects of education being greater in countries with higher levels of development, as in Latin America generally in comparison with Africa. Furthermore, these are reduced (but not eliminated) where there are effective family planning programmes, suggesting that

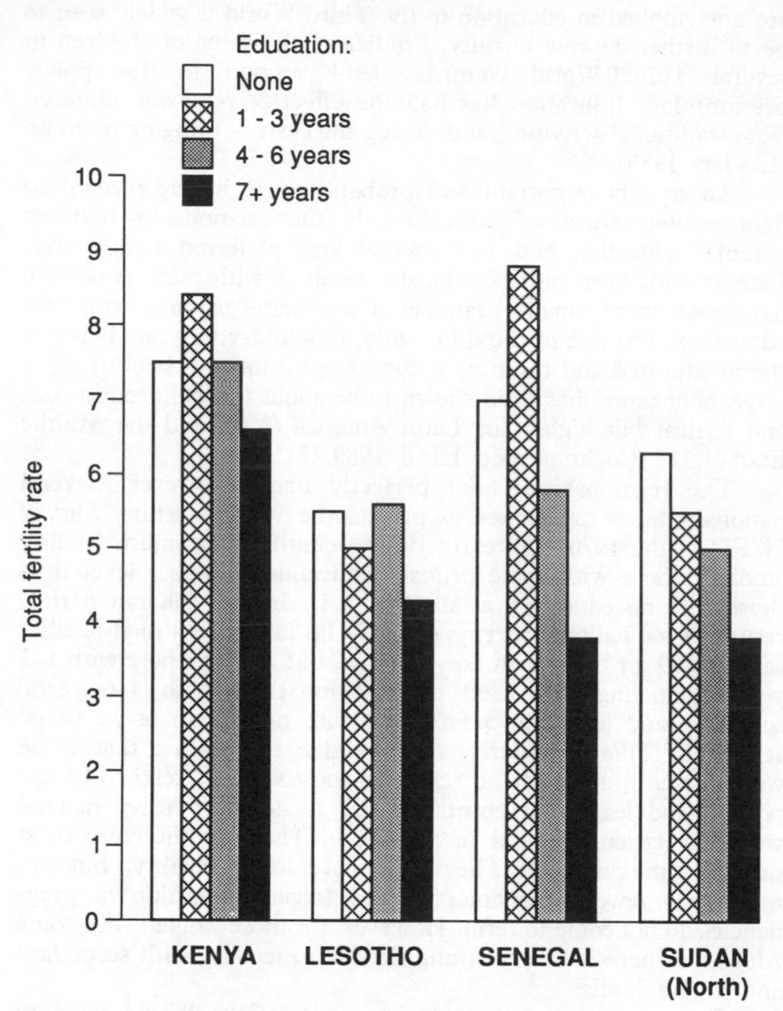

Fig. 5.1 Fertility and education in four African countries (from Awusabo-Asare 1988:8)

family planning programmes may hasten the use of contraception amongst those with less formal education. The education variable is inevitably associated with other socio-economic variables, and is reduced in importance as an explanatory variable when these other variables are standardized. Nevertheless education has been shown to be a more important determinant of fertility than region, rural–urban residence or occupation status for both males and females (Freedman 1987:784).

The relative importance of education in the context of other socio-economic variables has been confirmed by the Ghana study in the WFS round (Table 5.1). The range of values for younger as well as older married women of number of children ever born is much wider for the four education groups (ranging from 4.4 to 2.1 children) than for residence (a range of 3.8 to 3.2), region (a range of 4.1 to 3.0 for ten regions), or religion. Awusabo-Asare submitted the Ghana data to a multiple classification analysis and concluded that 'formal education is the variable with the strongest effect on completed fertility, followed by region' (Awusabo-Asare 1988:432).

As more detailed data become available, the observed

Table 5.1 Mean no. of children ever-born to ever-married females by age and socio-economic characteristics: Ghana 1977/78

Variables	Total	Aged 15–29	Aged 30–49
Education			
No schooling	4.4	2.1	5.6
Primary	3.4	1.9	5.4
Middle	2.4	1.8	4.5
Post-Middle	2.1	1.3	3.4
Residence			
Rural	3.8	2.0	5.5
Urban	3.5	1.9	5.2
Large Urban	3.2	1.7	4.8
Region			
Western	4.1	2.1	6.0
Central	3.9	2.0	5.6
Greater Accra	3.0	1.7	4.4
Eastern	4.0	2.0	5.7
Volta	3.9	2.1	5.3
Asante	3.4	1.8	5.4
Brong Ahafo	4.1	2.0	6.0
Northern	3.6	1.9	5.0
Upper	3.6	2.0	4.9
Religion			
Christian	3.5	1.9	5.4
Muslim	3.5	1.8	5.3
Others	4.0	2.1	5.4
Total	3.7	1.9	5.4

Source: Awusabo-Asare 1988.

complexity of the relationship between education and fertility grows, but the policy implications remain clear: in a majority of circumstances an enhancement of female enrolment rates, and especially in ensuring more girls stay on for much longer in school, will have a negative impact on fertility. This general finding provides a further strong justification for investment in educational expansion. As the World Bank Policy Paper, *Education in Sub-Saharan Africa. Policies for adjustment, revitalization and expansion* states:

> . . . there is a strong negative relationship between how much education a woman receives and the number of children she bears during her lifetime. Men and women with more education, in addition to having fewer children, tend to live healthier and longer lives. And numerous studies have shown that parents' education affects children's survival and enhances their physical and cognitive development.
>
> (World Bank 1988:7)

That this apparently valid general relationship should apply is a function of educated women having a proportionately higher social and, more particularly, economic status. They are therefore more likely to be able to make independent decisions about family size, or at least be much involved in joint decisions. As a result of their being in school they are likely to be married later, reducing the overall period of exposure to pregnancy. On the other hand, educated women in Third World countries, especially those in formal employment, are more likely to bottle-feed than to breast-feed their babies. They thus increase their exposure to pregnancy and have shorter intervals between births. However, this increased exposure is more than compensated for by a much increased knowledge of, more positive attitudes towards and more frequent and effective use of modern contraception.

Bolivia has a total fertility rate of 4.9, substantially higher than the Latin American average of 3.7, according to its Demographic and Health Survey data for 1989. This high rate is attributed to a relatively low contraceptive prevalence rate of 12 per cent among married women. However, when these data are differentiated by educational status clear differences in actual fertility, desired fertility and contraceptive prevalence rates are found. Women with nine or more years of education (30 per cent of women of reproductive age) have fewest children, want even fewer children and more than one-quarter of them use modern contraceptives (Table 5.2). Only 2.4 per cent of women with no education use contraceptives. For this group there is a very large gap of over two births per woman between actual and desired number of births.

One recent study in Khartoum, Sudan, has further illustrated the importance of education in contraceptive use, indicating important differences among the educated when disaggregated by socio-economic area of the city. Educated women in the highest

Table 5.2 Fertility and education in Bolivia

	Educational status (currently married women)		
	No education	6–8 years	9+ years
Actual fertility (births per woman)	6.1	4.5	2.9
Desired fertility (births per woman)	3.8	3.1	2.2
Modern contraceptive use (%)	2.4	16.8	25.7

Source: Bolivian Government 1990.

status areas know of and have access to contraceptive methods of various kinds, but are less likely to use them. As a result they have higher fertility than educated women in middle-status areas. Women in these areas have the lowest fertility of all women in this survey (El Dahab 1992). Social as well as economic incentives for contraception will result in different prevalence rates at any given level of education, but the differences between social areas of the town are substantially less than differences within each of these areas associated with educational status.

Educated women are most likely to be responsive to family planning media campaigns – in newspapers and other written material – and to relate personally to the nurses, midwives and other health professionals involved, themselves educated women. They are more likely than uneducated women to live in towns, or areas of towns with good access to health care and to contraceptive outlets, and are more likely to be able to afford to use the facilities that are available. Furthermore, educated women in Third World countries tend to be in paid employment. Though they are able to employ servants, or have other female relatives to look after their children, and though the same welfare benefits may be available (e.g. statutory maternity leave), there is a disincentive for working mothers to have large families.

Levels of fertility in the Third World are highly responsive to levels of education, both at the aggregate and individual scales. Mothers' education is especially critical. Enrolment expansions of the last decade are expected to be strongly associated with further falls in fertility, and as the expansion of girls' enrolments at the secondary level becomes more evident, the downward trend in fertility will be further supported. Education gives women more control over their own lives, and this is more evident in their fertility experience than in their participation rates in the labour market.

Education and mortality

Mortality, the other component of 'natural' population change, has fallen in recent years much more than fertility has fallen throughout the Third World. It is evident in countries which have experienced substantial fertility decline, as in Latin America, as well as in countries that have not experienced substantial fertility decline, as in many African countries. However, the fall in mortality in the Third World has been even more closely associated with education than has the fall in fertility. Thc negative relationship between education and mortality has been recognized at all scales in a wide range of recent studies, classically in the benchmark analysis for Nigeria by Caldwell (1979). It has been recognized in international comparisons of major world regions, e.g. by Hobcraft *et al.* (1984) using WFS data, within each of these world regions, at the national scale and at the local scale in inter-village differentials (Freedman 1987). At all these scales the findings seem to be generally consistent – as levels of education of males and, more especially, females rise, so overall rates of mortality fall.

Educated adults, those who themselves have been exposed to schooling, tend to have higher status jobs, to be better fed and live in better housing, to have better physical access to higher quality health care and a greater ability to pay for it. In general it is not surprising that the levels of adult mortality of the educated should be lower than those of the uneducated population. Athough data on adult mortality rates in Third World countries have, until recently, been poorly collected and only partially analysed, it has been estimated, using data from the WFS survey in Lesotho, that the parents of respondents with secondary schooling live six years longer that the parents of those respondents without any schooling (Timaeus 1984).

There are some anomalies that are associated with education, though not directly. In several African countries, and especially in urban areas, the largest single cause of death among young adults, a group with very low mortality rates, has been road accidents, and access to cars or other private transport has been biased towards the higher status groups, disproportionately well educated. However, the rapid increase in AIDS prevalence in Africa has affected the young adult population mostly, including a disproportionately large number of élite educated workers, at least in the earliest phases of the epidemic. The pattern of age-specific cause of death may sharply change as overall adult mortality rates rise (Barnett and Blaikie 1992).

Much more important for this discussion, however, as a proportion of all deaths and as a proportion affected by education, is the relationship with children's mortality. The main reason for the rapid, widespread and sustained fall in mortality, whether measured by falling crude death rates or rising life expectancies, has been the fall in mortality rates of the young. This has affected infant mortality rates (IMR = number of deaths of children aged less than one year as a proportion of all births in any given period) in particular, but also childhood mortality rates (CMR = number of deaths of children aged 1 to 4, inclusive, as a proportion of those aged 1 to 4 in any given period). Infant mortality rates were typically well in excess of 150/1000 in most Third World countries before this century, but could rise to 500 (i.e. one-half of all children dying in their first year of life) in years of crisis mortality, i.e. periods of major disease epidemics, famines or national disaster. Now they are generally below 100, even in Africa, the continent with highest IMRs. Child mortality rates are generally substantially lower than IMRs, and have fallen even more rapidly than IMRs in most countries, largely as a result of vaccination and immunization programmes, but they are still higher than IMRs in many West African states (Hill 1991).

There is more than simply an ecological correlation with aggregate educational expansion in recent decades, for levels of IMR and CMR are strongly differentiated by education status of parents of the children, both father and mother, but particularly the mother. Cleland and van Ginneken have exhaustively reviewed the effects of maternal education on childhood survival in developing countries, and have concluded that there is:

> a truly astonishing sensitivity of childhood survivalship in developing countries to the length of formal schooling of the mother. Even after adjustment for economic factors, one to three years of schooling is associated with a fall of 20 per cent in childhood risks of death and further large decreases are recorded with successive increments in educational attainment. This strong relationship is found in all major regions of the developing world, and persists in countries with accessible and effective health services, and those with weak primary health care systems.
>
> (Cleland and van Ginneken 1988:1365)

The extent of the decline in mortality with education seems greatest in Latin America, but they show it is strong also in the two population 'giants' of the Third World. In India, using data from a national survey of infant and child mortality in 1979, the index of mortality of children of literate mothers is 0.70, for those with more than primary education it is 0.49, when the figure for illiterate mothers is 1.00. Indian children are twice as likely to survive their first year of life if their mother has attended secondary school. In China, also, where the figure for illiterate mothers is 1.00, the index of mortality is 0.77 for mothers with elementary school education only, 0.48 for those with junior school education and 0.42 for those with senior school education and above.

The strength of the relationship seems to be different for different ages of children. Traditionally the mortality in the first year of life is divided into neo-natal mortality (deaths in the first month) and deaths in the rest of the period. In that first month, deaths are due almost entirely to biological or congenital problems, and not influenced so much by mother's behaviour or the environmental conditions of the home. Beyond this very early period the role of the mother and of the broader environment of the child becomes more prominent and education then begins to exercise its effect. In Ghana, for example, data on infant and childhood mortality are available from the Demographic and Health Survey of 1989, and these offer a useful complement to similar data on fertility and fertility differentials in Ghana discussed earlier (Table 5.1). A calculation of the infant and child mortality rates for Ghana as a whole has the IMR and CMR roughly similar. Eight per cent of all children die in their first year and 15 per cent before their fifth birthday, but these proportions rise to 14 per cent and 22 per cent, respectively, for some regions of the country (Table 5.3). These figures are below the average for Africa as a whole. The rates for Ghana are differentiated by rural–urban residence and by region, but differentials by education again show the largest range, particularly for child mortality. While IMRs by education of mother cover a fairly small range, and are higher for mothers with secondary education than for mothers with middle education, the CMRs display a much greater variation: children born to mothers with no education are four times as likely to die aged 1–4 than are children born to mothers with higher education. One major factor reducing childhood mortality has been immunization against measles, whooping cough and diphtheria, administered normally between 9 months and 1 year. The differential uptake of vaccination by education of the mother may be a principal factor at work in Ghana to produce such large differentials for this age group.

One surprising and certainly controversial theme in the mortality literature is the effect of mother's education on the differential mortality of male and female children. In northern India and Bangladesh male:female differentials in childhood survival are well established as boys are given preferential treatment in food and medical care over girls, for reasons that reflect the broader social values ascribed to males and females in these areas. What is the effect of mother's education on the differential? Bourne and Walker (1991) argue, using data for India, that the effect of mothers' education has been to narrow the male:female differential in survival, with an effect that is greater than the effect of rural or urban residence. The average education effect was to reduce deaths of boys in northern India by 77 deaths

Table 5.3 Infant and childhood mortality by socio-economic characteristics: Ghana, 1988 (Rates per thousand)

Background characteristic	Infant mortality rate (< 1 year) 1978–87	Childhood mortality rate (1–4 years) 1978–87	Mortality rate (0–5) 1978–87
Residence			
Urban	66.9	68.8	131.1
Rural	86.8	82.9	162.5
Level of education			
No education	87.7	95.2	174.6
Primary	84.8	68.5	147.6
Middle	69.7	64.0	129.2
Higher	79.1	22.2	99.5
Region			
Western	76.9	80.4	151.2
Central	138.3	81.9	208.8
Greater Accra	57.7	48.9	103.8
Eastern	70.1	73.2	138.1
Volta	73.5	63.8	132.7
Ashanti	69.8	80.0	144.2
Brong Ahafo	65.0	61.6	122.6
Upper West, Upper East, Northern	103.1	132.3	221.8
Total	81.3	78.9	153.8

Source: Ghana Government 1989.

per 1000, but for girls this reduction was nearly 100 per 1000. This they argue is due to mothers being aware of and seeking to redress the general social bias against girls by discriminating against them rather less in traditional food allocation and other life chances than illiterate mothers do. By contrast, Bhuiya and Streatfield (1991) come to the opposite conclusion in a more detailed study in rural Bangladesh using a follow-up of nearly 8000 births during 1982. As in India maternal education improved the children's survival overall, but change in mother's education from no schooling to 1–5 years in school resulted in a fall of 45 per cent in predicted risk of death for boys, but only 7 per cent for girls. For levels of mothers' schooling above six years the corresponding falls were 70 per cent for boys and 32 per cent for girls. They attribute this to women's education acting to reinforce the social structures, sensitizing women even more strongly to the differential social and economic importance of male and female children at the prevailing low levels of development in the country. Clearly here is a theme of importance that warrants further study in a range of cultural and economic contexts, for not only does the theme provide important findings in its own right. It can also shed light on the more general processes of how education acts on the one hand as a force for social reproduction (as might be argued from the Bangladesh findings) or on the other for social transformation (as in the Indian case).

The educational differential is apparent with changes over time, but the gap between groups seems to be narrowing in countries at low levels of development and at high levels of mortality that are falling quickly. Cleland and van Ginneken cite the case of China, 1964–76, to illustrate the trend in the education differentials. The low mortality level of children of parents with senior schooling remained roughly constant over the period, at about 20/1000. In contrast the rate for children of illiterate parents fell by about one-third, from nearly 90/1000 to 60/1000 in this same period, and overall the levels of mortality narrowed but remained consistently as expected in the general education/mortality relationship (Fig. 5.2). Werner's study of Kenyan childhood mortality (deaths before age 5) using census and other data revealed a similar but rather more complicated picture. In that case overall levels of mortality fell by 36 per cent between 1954 and 1977, but for the two most educated groups, those with primary and secondary education, levels actually rose in the 1970s, while the general trend

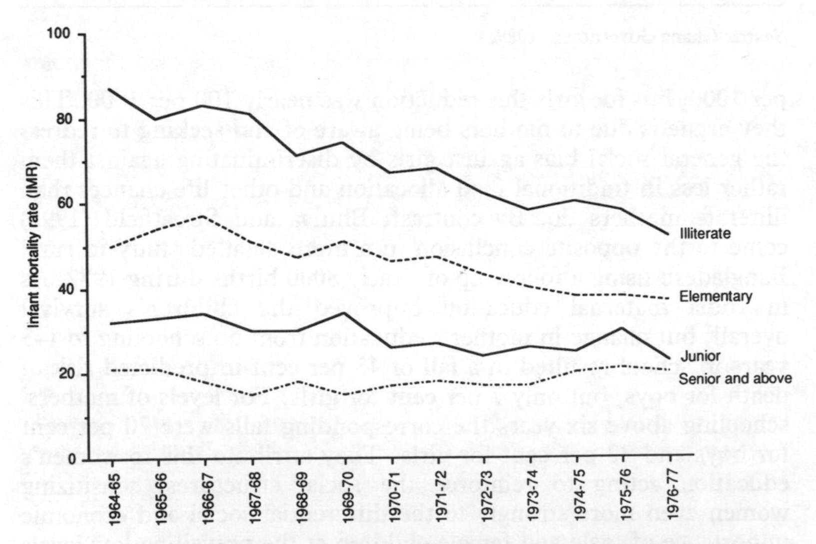

Fig. 5.2 Infant mortality rate by mothers' education: China 1964–77 (from Cleland and van Ginneken 1988:1364)

was for decline amongst those mothers with no education. She attributes the small rise in childhood mortality amongst the educated as an indicator of the massive extent of educational expansion in Kenya in this period, to the extent that education brought with it fewer and fewer economic rewards. For any given level of education the actual income earned, and more generally the quality of life, was consistently falling (Werner 1990).

Some of the possible problems of separately identifying the effects of education on changes in mortality over time are reflected in a comparison of DHS data for Kenya, 1989; and Ghana, 1988 (Table 5.4). We have already seen how the IMRs in Ghana are differentiated by educational status (Table 5.3) and this is paralleled for Kenya, where the IMRs are 72/1000 for children of mothers with no education, 59/1000 for those with some primary education, 49/1000 for those with completed primary education and 42/1000 for those with secondary education or more. Overall levels of mortality are much higher in Ghana than in Kenya, but in Kenya the child mortality rate has fallen most sharply to be about half the IMR for the 1984–89 period. In Ghana it had fallen sharply between 1973–77 and 1978–82, but rose again, 1983–87, in a period of severe economic and environmental stress in which health services virtually collapsed. In this period the child mortality rate rose, mainly since, as we have argued, it is highly sensitive to immunization. But the IMR continued to fall in this period. In Kenya, by contrast, there was no such severe economic or environmental shock, yet IMR rose between the two later periods, perhaps as a result of the education expansion factor identified by Werner (1990). The child mortality rate continued to fall and the overall 0–4 rate was less for 1984–89 than for 1979–83, but not by very much.

Table 5.4 Mortality indices: Kenya and Ghana, 1973–87 (Rates per thousand)

	Kenya			Ghana		
	1974–78	1979–83	1984–89	1973–77	1978–82	1983–87
Infant mortality	64	58	60	100	86	77
Child mortality	44	38	32	97	72	84
Under 5 mortality	106	93	89	187	152	155

Source: Kenya Government 1989a; Ghana Government 1989.

Cleland and van Ginneken (1988) identify the range of complementary explanations for the near universally observed negative relationship between maternal education and mortality. One set of explanations sees the importance of education crudely in terms of income and socio-economic status. Education in this sense acts on the health and mortality of children mainly indirectly, since it allows the families of educated mothers to be better off, have better housing and nutrition for their children and to be able to afford better access to health care when necessary. Education has no independent effect on the children, but only operates though a broad range of associated socio-economic variables. Increasing income of the family (as of society in general) will be sufficient to secure falling mortality.

The role of education may be more direct in that it can allow parents better exposure to information and knowledge. Education can make parents more aware of basic causes of ill-health and of means to prevent disease and infection. They will have a better understanding of the need for hygiene in the home, and of medicines to treat illness. They will be able to read instructions as well as be more susceptible to media and neighbours' information and advice. The educated will be better disposed to 'scientific' medical care for the sick and less dependent on traditional remedies and healers or on spiritual or other non-objectively rational explanations of disease, and generally will 'replace the more resigned and fatalistic view of the uneducated mother' (Cleland and van Ginneken 1988:1364).

This enhanced predisposition to a positive view of the use of medicines and health care is most effective where the means are available to use that knowledge and attitude in the form of an accessible and effective health service. The effect of education is likely to be greatest where health interventions can be made available to children and used by educated parents. The case of higher rates of immunization amongst children of educated parents, cited above as being particularly instrumental in lowering childhood mortality rates, is especially appropriate here, but this is only the most obvious example of general acceptance and use of Western remedies and health care. These are more likely to be used at an appropriate time, as a first resort when a child is ill rather than as a second or third best after other remedies have been seen not to be effective.

It would appear that the greatest effects come where high levels of education are found in areas with a good health care delivery system. Caldwell (1986) argues that it is in such areas that the most dramatic justification for the force of the education factor in demographic change can be found. He reviews the general argument amongst demographers over the relative importance of health interventions and economic development as the mainsprings of mortality decline in the Third World, as indeed has been argued for nineteenth-century Europe and North America. A statistical comparison of data on 99 Third World countries led him to the conclusion that the main difference between countries with substantial achievements in reducing mortality in the years since the Second World War, countries such as Sri Lanka, China, Burma, Jamaica and Kerala in India, are that these are areas with high levels of mass education, and thus high proportions of girls in school. Female primary school enrolment is a sensitive indicator of levels of mortality and the extent of mortality decline. It is more critical than per capita income or the numbers of physicians or nurses per head or per capita calorie intakes.

Caldwell has argued for the notion of the 'Health Transition' to help explain mortality change in terms of health care and public health measures rather than in terms of development, as expressed in rising incomes or social change, as the critical factor. However, his analysis recognizes the synergy between efforts in education and efforts in health. The mortality declines have come about where countries or regions have expressed a positive political will to allocate resources to social expenditure, notably education and health, rather than directly to economically productive expenditure. These allocations have not been done primarily to achieve demographic objectives: they were initially designed to raise the quality of life of the populations as a whole. However, mortality decline has been a major benefit. Kerala offers the most obvious case. It is one of the poorest states of India, yet has high rates of enrolments of boys and girls in school and has a well-established and well-spread health care system. Already by the 1970s it had the lowest mortality rates of all the major Indian states, and its performance has continued to improve more rapidly than the country as a whole. Its crude death rate fell 27 per cent, from 8.9/1000 in 1971–73, to 6.5/1000 in 1983–85, at a time when the Indian rate fell only 24 per cent, from 15.9/1000 to 12.1/1000 over the same period (Krishan 1989).

Education and the demographic transition

Yet Kerala has also experienced very substantial fertility declines in this period. Crude birth rate fell 23 per cent, from 30.5/1000 in 1971–73, to 23.6/1000 in 1983–85, while the national fall in this period was only 8 per cent (from 36.3/1000 to 33.5/1000). Taking fertility and mortality together, Kerala's rate of natural increase had fallen from 2.2 per cent per annum to 1.7 per cent per annum.

Its low birth and death rates suggest that it has progressed furthest of all the Indian states along the path of the demographic transition (Caldwell *et al.* 1985; Krishan 1989). This general model of population change hypothesizes a regular experience of change in populations from a low rate of growth with high rates of fertility and mortality (the pattern of pre-industrial societies), through a period of growing rates of growth caused by falling mortality while fertility is constant (the pattern evident in much of the Third World, 1950–80), into a period of fairly constant mortality but falling fertility in which rates of overall population growth are falling (as in East Asia and Latin America at the present time), until a final low equilibrium stage (characteristic of contemporary developed countries) is reached. While most Indian states have experienced only small declines in fertility but more substantial falls in mortality, and therefore have annual population growth rates in excess of 1.5 per cent, Kerala seems to have progressed to nearer the low equilibrium position with much smaller differences between fertility and mortality. Why is Kerala different from the rest of India? Given the discussion of mortality and education in the previous section, what has been the role of education, in Kerala but by implication more generally, in accelerating the apparent transition from high to low overall population growth?

The demographic transition model is an empirical generalization, based on the patterns of population change experienced by those countries that have completed the transition – i.e. in North America, Europe, Australasia and Japan. It describes the process of change but can only infer the causes of the changes and the causal relationships between changes in fertility and changes in mortality. It implicitly assumes fertility to be a lagged response to mortality, that factors responsible for declines in the death rates will be systematically linked to factors subsequently provoking changes in fertility (Caldwell 1991; Woods 1982). These ideas are not entirely convincing. Changes in mortality are often due to the largely macro-structural changes in the economy and society, whether it be improvements in levels of development and incomes, or improvements in public health and medical advances, as implied by the Health Transition School, including Caldwell (1986). Dying is not normally a matter of individual choice. Fertility, by contrast, is the result of decisions and choices of individuals. It is therefore more likely to be affected by cultural and social values and changes, such as changes in marriage and attitudes towards marriage that operate at the micro-scale. It is less likely to be directly affected by broad structural changes such as rising incomes, though these will have an indirect effect.

It has been shown in this chapter that education is a critical variable for both fertility and mortality. Both are sensitive to education differentials, and these operate in the same direction. More education seems to have the effect of lowering levels of mortality, especially infant and child mortality, and also has the effect of lowering levels of fertility. In both cases the differential effects of education are greater for women than for men. In this respect the situation in the Third World is rather different from that of nineteenth-century Europe and North America where the fall in mortality generally began before mass education had been implemented, and in the majority of countries (the obvious exception being Great Britain) before massive urbanization or industrialization. The fall was largely a function of public health improvements and medical advances, with education having only a minor role. Falls in fertility began rather later, perhaps 50 years later, by which time education had advanced, as had other social changes, to create the behaviourial and cultural conditions necessary to alter family size.

In the Third World, however, the rapid expansion of education to mass education levels has occurred at much the same time as the substantial falls in mortality (i.e. 1950–80), and rather before the main period of fertility decline. This is the case even for those countries that have experienced fertility declines, mainly since the 1970s as in the Philippines (Hackenburg and Magalit 1985), and, as in Africa, has certainly preceded major fertility falls. Even here, or more properly *particularly* here where there is high overall fertility, education differentials are most marked for both fertility and mortality, and there is a presumption that more education for more people for longer, and especially for girls, will result in fertility levels moving from those of the 'no education' or 'little education' groups to those currently in the higher education groups. However, as Werner's analysis of the Kenyan data has indicated, as education is more generalized its differential impact is reduced, though it is by no means eliminated. For any given level of education its effect is less. The effect of further expansion in education is likely to be less than earlier provision seems to have been.

Another general difficulty associated with the predictive capability of the demographic transition model concerns the variable levels of educational enrolment at which the decline in fertility and mortality seem to be triggered. In nineteenth-century Europe, as noted above, enrolment rates were still low when mortality fell, and current levels of enrolment in most Third World countries are higher than they were in mid-nineteenth-century Europe (Caldwell 1991). Similarly, levels of enrolment in some African countries (e.g. Kenya, Zimbabwe, Ghana, Cameroun) are currently higher than they were in some East Asian or Latin American countries when they began to experience fertility decline.

There seems to be no universally valid enrolment threshold that needs to be reached to act as a trigger for demographic change. Education is clearly not the only factor involved. Even if it were, differences in the quality and impact of education associated with any given formal quantity of education, as measured by years in school, would result in a range of values at which the impact of education would begin to be felt.

It is likely, in particular, that population education will be able to have a substantial impact in those countries with already mass education and an effective direct population policy of family planning. The experience of Indonesia, Thailand and Philippines, for example, is that the introduction of population education into the secondary school curriculum reinforces the broader policy innovations associated with the better provision of family planning advice and supply and use of contraceptives. Population education would also be likely to affect mortality through maternal and child health components of the curriculum programmes. Many countries have introduced population education into the school curriculum, some only half-heartedly, others with no obvious effect on population change, but such changes within the school can clearly reinforce broader behavioural changes that the more general educational experience can make (*Population Reports* 1982, 1989).

Education can provide an integrating link between the causes of mortality decline and of fertility decline. In so doing it gives greater logical cohesion and predictive credibility to the demographic transition model. Caldwell's (1986) case for mass education being a major condition for mortality decline applies also to fertility decline, for he also argues strongly for the role of education as a force for fertility change (Caldwell 1980). In particular he argues that fertility declines are most likely where there is a restructuring of flows of wealth in society, away from traditional upward flows, where the young create wealth to support the older generation. In these circumstances it will be 'rational' to have large families. High fertility will ensure that parents will have many 'hands to work' and children are a security for their comfortable old age. However, where the cost of children is high and educating them is expensive, family size will fall as the flows of wealth will be towards these children in school, and into their own marriages after they have left school. The educated, especially in difficult economic times, will be more likely to prefer nuclear rather than extended family relationships, with less of their income allocated to supporting the older generations. Education is one means of accelerating the necessary structural social changes that he argues are needed for fertility reduction. Education, in this sense, is constructed as a factor for social transformation rather than social stability.

The expansions in education that were outlined in previous

chapters find a further justification in their effect on population growth. Educational expansions are of course to be justified primarily in their own right – to provide better life chances for these who go to school and to promote the economic well-being of society as a whole, and cannot be justified primarily in terms of demographic impact. This is particularly the case for needs in improving educational opportunities for girls. But with disproportionate expansion of girls' education the demographic impact will be greater, an important contribution to the broader context of improving the quality of life and status of women in the Third World (Simmons 1988: esp. Ch.7).

Chapter 6

Human resource development

Education makes individuals more productive in an economic sense. They are more able to contribute to the development of the local and national economy. That the economic contribution of an educated person is normally greater than that of an uneducated person is evident in the public and private demand for education, as discussed in Chapter 1, and in the income rewards to educated people. Since education is widely taken to be a prime mechanism for economic development in any country in the Third World, it is to an economic perspective on education, schooling and training that this chapter turns. It considers how different approaches to the education and training of the population can affect the nature and pace of development, and develops two rather different examples, from Pakistan and Kenya, to substantiate the argument in the broader context of human resource development.

People acquire different skills and abilities in the course of their lifetime. These skills are not all acquired in schools or even elsewhere in the formal education system, but the system clearly enhances the 'quality' of the population. 'Quality' is used here in an economic rather than a eugenic sense. The skills of the people enhance the amount and use of resources available to society, and are therefore important factors in development. Population is itself a resource. To conceptualize population in this way as a resource is difficult, especially for those who have themselves seen education as an individual benefit that is broader than the purely economic. However, this conceptualization of population as a resource has been of growing significance in development studies in recent years, and notably in the growing prominence of the term, *human resource development* (HRD). HRD has become a central objective of and means to development. Formal education and training policies designed to enhance the quality of the human resource base are critical in any particular development context, and can vary very considerably amongst the countries of the Third World.

A discussion of HRD offers a further opportunity for the analysis of important education and development interactions. The first two sections of this chapter raise some conceptual, policy and planning issues for HRD and how they are affected by the education system. In the third and fourth sections these general issues are examined in two specific and quite different development contexts – Northern Areas of Pakistan and Western Province, Kenya – to illustrate the flexibility of HRD as an operational concept, and its role in educational and training programmes for the general development strategies currently being applied in these areas.

Education and human resource development

There is a general reluctance amongst social scientists to consider 'people' in terms normally used to measure cattle or iron ore, as units of value or work, mindless automatons in an economy or society over which they have no control. Yet the terms 'human resources' and 'human resource development' are familiar in the development studies literature, and their use has been increasing in recent years. This increased prominence of HRD has been associated with new ideologies in economic analysis, styled by Toye (1987) as a 'counter-revolution in development theory and policy'. Specifically, the New Right has moved to centre-stage in economic thinking – in the Third World as in the developed world and led by such organizations as the International Monetary Fund and the World Bank – and has introduced new ways of thinking about the role of individuals and their skills in the development process.

The Malthusian calculus of classical population analysis explores the relationship between population and resources in quantitative terms: numbers in the case of population; and food and land in the case of resources. While Malthus would allow for some improvements to the land resource (use of fertilizer or new technologies) to maintain food supply, his calculus precludes the possibility of seeing people other than as consumers and as producers in a fixed ratio with land and food. This view extended into the mainstream of classical economics, and has dominated it throughout the nineteenth century and for most of the twentieth century. Though Alfred Marshall some 100 years after Malthus could write that '[knowledge] is the most powerful engine of production; it enables us to subdue nature and satisfy our wants' (quoted in Schultz 1981:17), it was not until the 1960s that human resource development became an important issue for development theorists. Then it came from the political Right as a reaction against the Keynesian dominance of the post-Second World War period, and reached its greatest academic prominence in Theodore Schultz's Nobel Prize address of 1979, 'On the economics of being poor' (Schultz 1981:3–58). At a more practical and directly policy-relevant level it formed the core of the World Bank's *World Development Report, 1980*, which addressed the theme of *Human resources and development*.

Schultz argued that 'the main thrust of my argument is that the investment in population quality and in knowledge in large part determines the future of mankind. When these investments are taken into account forebodings concerning the depletion of the earth's resources must be rejected' (Schultz 1981:xi), and that 'the decisive factors of production in improving the welfare of poor people are not space, energy, cropland; the decisive factors are the improvement in population quality and advances in knowledge' (Schultz 1981:4). These seem to be surprising conclusions from a scholar whose main work had been in agricultural economics, but who concluded that even for agriculture the land resource question was of lesser general significance than the human resource question. It did, however, fit well with Julian Simon's more high profile view on *The ultimate resource* (1981). Together their views effectively won the day at the 1984 World Population Conference in Mexico. The need for reductions in highest rates of population growth was conceded, even by governments of the highest growth countries which had previously been reluctant to admit to the need for direct population policies, especially in Africa. In addition there was also a recognition of the need for a renewed emphasis on population quality. Investments in education are universally seen as a prime mechanism to enhance the resource base.

The impetus for ensuring such a shift in resource allocation has been provided in practice by the international community, and particularly by the World Bank. It had lent for educational development projects since 1962, and by the late 1970s, at the height of its concern for development with redistribution of wealth in its basic needs thrust, it was heavily targeting the rural poor for schools, dispensaries, etc., that were held to be both equitable and also developmentally justified, in that they would provide a platform from which the poor would be able to raise their own productivity and incomes. Investing in population quality was necessary for these improvements, though the term 'human resource development' was not greatly used at the time. During the course of the 1980s, however, as explicit concern for equity faded in favour of efficiency, there was a clear shift in investment priorities towards human resource development away from investment in physical capital; away from bricks and mortar towards people directly. This was a shift that the World Bank could justify on purely economic criteria, as has been argued in earlier chapters, but it had its equity effects too. The shift is justified in education in various ways. At the micro-level it was supported by the consistent findings of high returns to educated individuals as measured by their current and lifetime earnings. At the macro-level it was confirmed by measures of the overall effect of educational investments in the national accounts: 'The overall conclusion is clear; increased education of the labour force appears to explain a substantial part of the growth of output in both developed and developing countries since 1950' (Psacharopoulos and Woodall 1985:17). In many recent policy documents on education, and also more generally, there is an ideologically consistent agenda that gives to human resource development priority in development strategies for the 1990s.

It is clear in general terms what is meant in these development strategies by 'human resource development': defined negatively it can be anything that is not investment in bricks and mortar; more positively, it requires expenditure that is targeted to people directly to raise the actual and potential economic productivity of the population, identified as 'quality'. It seeks to raise the productivity of the resource, to enhance its ability initially to generate and eventually to sustain rising incomes. It requires investments that allow people to raise their own income (either from self-employment or in the labour market) and, where appropriate, the income of their employer through more or more valuable output. Clearly health improvements are appropriately included here as investments in human resources, justified by economic as well as humanitarian criteria, as in the WHO objective of 'Health care for all by 2000'. Healthy people are more productive. They also have higher levels of personal non-economic satisfaction and contribute more fully to the life of their communities. So too, it is assumed, with educated people.

In education the means and objectives of raising population quality are much more controversial than they are in the health sector. They invoke a large number of debates about appropriate measures of education, type of curriculum, length of schooling, etc. Education cannot, in other words, mean an open-ended commitment to more education of whatever type. It must be premissed on an ability to measure the benefits of particular types of investments to be set against the expected benefits of other investments in the education sector. Issues of measurement and evaluation of the range of alternatives are the source of major controversies in education and development. Particularly relevant to current policy issues in many countries and with major implications for the course of population change is the issue of choosing between in-school education on the one hand and out-of-school 'training' on the other as alternative or complementary paths to human resource development. Under what conditions and in which respects might in-school, formal education be a preferred strategy to on-the-job training for enhancing the skills and productivity of a population?

A recent policy paper of the World Bank on *Vocational and technical education and training* (World Bank 1991b) has come down firmly on the side of a 'training' strategy that emphasizes the out-of-school private sector, on-the-job programmes. It sees the role of formal schooling as being necessary to provide the basic preconditions for more specific training, with a broad base of students exposed for perhaps nine years to a general diversified curriculum. However, it argues that specific vocational and technical training streams in secondary schools are not only expensive, but also they do not normally greatly enhance real competences of those who have been exposed to the in-school training compared with those who have been exposed to a more general curriculum. Better that skills are learnt in a context where they have real meaning and direct application. Secondary education can have a broad skill-based curriculum with practical work in basic science and technology, but not in specific job-related training. This is best left to the market, to employers who can focus the training needs to their own requirements, and also will control the supply to prevent an excess of poorly trained school leavers seeking work in a limited job market.

There has been a general over-supply of school leavers from vocational and technical courses in a great many countries. A common reaction of governments to the growing problem of educated unemployment has been to have more vocational education, on the assumption that with more skills more jobs can be created. This, unfortunately, has proved to be counter-productive for it has generally exacerbated the problem as even more school leavers come on to the skills market. Internationally comparative studies undetaken by the International Labour Office have shown how the educated consistently have higher levels of formal unemployment than the uneducated. In Indonesia, for example, a 1971 survey showed that the overall unemployment rate of those with no schooling was 7.8 per cent, but the comparable rate was 14.3 per cent amongst workers who had some education, with 9.8 per cent amongst those who had completed primary school, 14.5 per cent amongst those with secondary school certificates and 12.4 per cent amongst those with university degrees (Leonor 1985:225). A comparative study of unemployment, schooling and training in Tanzania, Egypt, Philippines and Indonesia concluded that expansion of vocational education in each case exacerbated the educated unemployment problem, and that:

> . . . vocational school graduates tend to have longer job waiting times than general secondary school graduates. Besides, the vocational school graduates tend to have lower earnings on average, perhaps not only because of occupational inflexibility but also for socio-economic reasons, the chief one being the disproportionately large number of financially poor students in

vocational schools whose social connections and information networks to good jobs may not be well developed.

(Leonor 1985:268)

There is a suggestion therefore that the problems of vocational education are not all, or even mainly, in-school problems, and the World Bank critique that seems to place the policy problem as one of appropriate institutional structures for training may ignore its social dimension. Futhermore, the World Bank's conclusions need to be set beside its wider ideological agenda of moving the focus of economic activity away from the state to the private sector. Lauglo (1992) argues, on the basis of his experience in vocational education in a number of African countries, that while the World Bank strategy would certainly be to the benefit of private sector companies, and particularly the more technologically advanced, often foreign-owned companies, it would offer only limited and company-specific training for a few rather than broadening the technological skill base of the labour force as a whole. The public sector in weak economies should not abandon its commitment to and direct involvement in skill training where skills are in short supply.

This controversy links to wider and more long-standing debates about the role of infrastructure and social overhead capital (SOC) relative to the role of directly productive activities (DPA) in the development process. The traditional view sees education as part of the wider infrastructure, an enabling factor for development, 'fertile ground without which developmental activities will not root' (World Bank 1988:21). Training, however, is more directly linked to production and can be seen to be a significant contributor to directly productive activities. They are usually to be preferred to SOC in the short term on the assumption that the weakness of infrastructure, i.e. the shortage of educated people and formal school provision, can be remedied by the market and by individual investment in training. In these terms, viable and efficient development comes from the stick rather that the carrot, from inadequacies or shortages in the human resource base rather than from its relative strength or abundance.

Policy interventions to raise the quality of the human resource will affect both the numbers and distribution of people. Investments in health, whether curative or preventive, clearly have a direct effect on mortality (as discussed in Chapter 5), and public health improvement is widely recognized as a major contributor to the substantial declines experienced by most countries of the Third World. Healthy people not only work better and more productively, but their life expectancies are higher. They experience lower mortality and fertility, and are more mobile (as will be discussed in greater detail in Chapter 7). Expansion of education and training opportunities, especially for the young, has further raised their propensities for migration, more so with higher levels of achievement and more specialist training. Here again, however, there is controversy over the extent to which the curriculum in schools or the type of training can effect the rate, distance and direction of movements in any population.

Manpower planning and the diploma disease

The most commonly applied approach to human resource development at the higher levels has been through manpower planning techniques. These have sought to estimate demand for skilled workers at some future date, and then to ensure that there is a supply of trained manpower to meet that demand. They were most prominent as a tool in educational planning in the 1960s and 1970s as a guide to planning the rate and type of expansions of the education systems to ensure a throughput of enrolments to generate a given number of engineers, doctors or other skilled manpower. A major controversy in educational planning at the time was over the alternatives of a social demand approach (based on the demands of the population) and a manpower approach (based on the demands of the economy) (Blaug 1968). Many countries had planned the expansion of the education system on the basis of optimistic assumptions of economic growth and a fairly rigid review of formal requirements of training programmes for skilled workers. By the mid-1970s it had become clear, at least in the majority of Third World countries and most particularly in Africa (Jolly and Colclough 1972), that a manpower planning approach was seriously flawed. Not only was it almost impossible to estimate the future demand for the overall labour force with any reasonable degree of accuracy (and most projections of demand made in the 1960s were wildly optimistic), but technological and social change added major complications to attempts to disaggregate the overall estimates. In practice the strength of social demand proved to be much stronger than the need for manpower planning, and its power was more a function of the political weakness of Third World governments in the face of massive population demand for schooling than of the technical difficulties of forecasting skilled manpower needs (Youdi and Hinchliffe 1985). This was even true in socialist countries or those, such as India (Verma 1985), where the practice of development had relied on centralized planning and a series of five year plans.

The failure of manpower planning to provide an adequate

match between the output of the education system and the demands of the economy meant that in practice there seemed to be an over-supply of educated manpower, since schools systems were driven by social rather than economic demand (Bray 1986; Lee 1988). As a result educated unemployment became a familiar feature in Third World countries. The educated were unable to find jobs in sufficient numbers that allowed them to use the education they had acquired, often at great expense and with great effort. Economies faltered, but the continuously expanding education system kept producing ever larger cohorts of formally trained graduates or secondary school leavers. There was continuous enhancement of the human resource base but these resources could not be channelled adequately to productive use.

This resulted in what Ronald Dore graphically described as 'the diploma disease' (Dore 1976). Experiences of school leavers with their formal qualifications – at degree or secondary school levels or whatever – meant that for any given level of formal qualification there was a continuous down-grading of jobs available to them. Conversely, for any given level of job there was an escalation of minimum qualifications needed for applying. Employers used educational qualifications as filters rather than certificates of real competence for any job. Government clerks, once employed as primary school leavers, now needed good grades in a secondary school or even degree examination. There was increasing reliance on formal qualifications rather than real competence – the so-called paper qualification syndrome (PQS). The process is described in Fig. 6.1. Increasing use of formal certification by employers creates additional demand for schooling, and especially at higher and secondary levels. Quality of education in these circumstances is likely to fall for any given level of education, but the number of

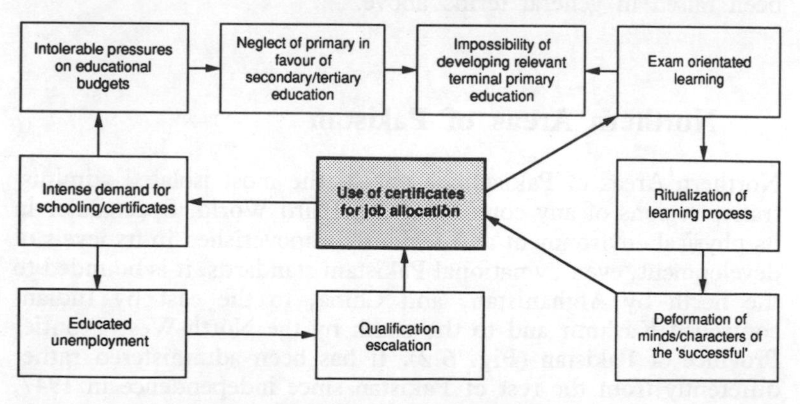

Fig. 6.1 The diploma disease (from Dore 1976:141)

people with the required certification rises. Employers are then able to be more selective and raise the minimum qualification threshold. All this results in greater reliance on formal examinations at the expense of 'learning' in the school curriculum. It also promotes the role of the education system as a pre-qualification for entry to the formal jobs market rather than as a primary vehicle for HRD in its strict sense.

In particular it further reinforces the vocational school fallacy, previously discussed in Chapter 4. Rural schooling, regardless of its curriculum content and especially where it seeks to provide those enrolled with skills that will be useful to them in rural areas, becomes a first step to an urban job. Going to school and achieving some level of formal qualification is a necessary first step in the search for a formal sector position. The link between education and migration is therefore integral to the operation of the diploma disease and the PQS. The formal education system as a primary means to human resource development needs to be carefully qualified in the light of particular circumstances. Dore suggested that the way forward to reduce the excessive dependence on formal and non-vocational qualifications is to have more on-the-job training, to encourage more self-employment, and more generally to broaden the scope of schooling from something undertaken by young people for a few years to something more akin to a lifelong education. All of these imply a shift from formal schooling to non-formal methods.

The examples discussed in the following sections of this chapter deal with marginal regions of two Third World countries that have experienced considerable attention in recent years directed to human resource development in the form of education and training. These case studies provide specific illustration of some of the issues and alternative strategies for HRD that have been raised in general terms above.

Northern Areas of Pakistan

Northern Areas of Pakistan is one of the most isolated administrative regions of any country of the Third World. Spectacular in its physical environment and seriously impoverished in its levels of development, even by national Pakistani standards, it is bounded to the north by Afghanistan, and China, to the east by Indian-controlled Kashmir and to the south by the North-West Frontier Province of Pakistan (Fig. 6.2). It has been administered rather differently from the rest of Pakistan since independence in 1947, with dircct control by the Federal Government rather than as a

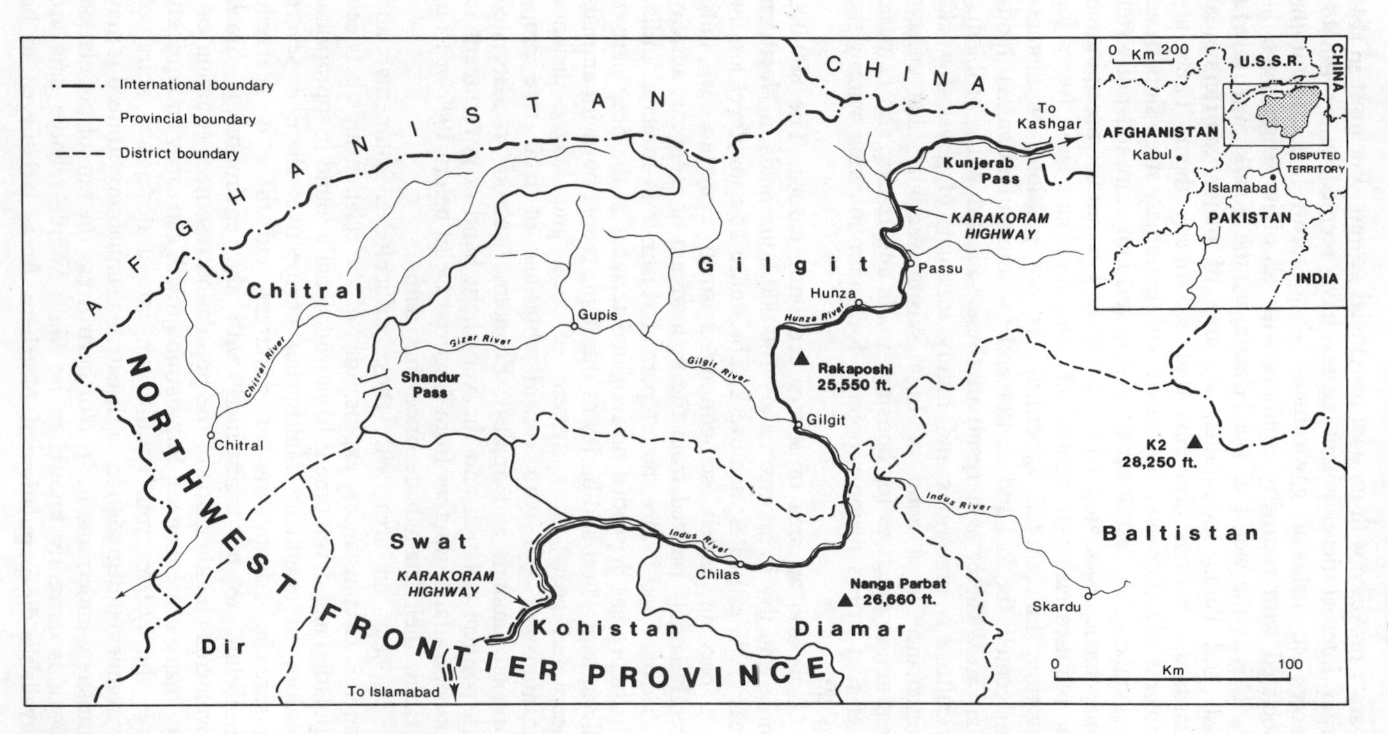

Fig. 6.2 Northern Areas, Pakistan

separate province with its own provincial powers. For most of that period internal development issues have seemed to be of less concern to national government than security considerations associated with recurring conflicts with all of its neighbours.

This is the world of the Himalayas, the Karakoram, Hindu Kush and Pamir mountains, an area of major environmental hazards and serious problems that accompany them. There are problems of soil erosion associated with extremely steep slopes and spectacular relief, fast flowing torrents and immature soils; of water management, associated with extremely low annual precipitation that requires elaborate control of snow-melt and large rivers for irrigation, for rain-fed agriculture is not possible; of biomass management for fuel and shelter and for human and animal foods in an area where, as a result of these soil and water problems, agriculture is precarious and highly seasonal; of physical works maintenance in a zone of high seismic activity that affects construction practices for housing, roads and tracks; and of basic relief that brings major problems for communication within the region.

It is also an area of severe economic hazard. The 800 000 people who live in approximately 100 000 households in Northern Areas (1988 estimates) achieve a subsistence at a low level for the most part in largely self-contained small communities, until recently largely isolated from the mainstream of economic activity in Pakistan, and before that as part of Kashmir in Imperial India. Subsistence has depended on irrigated terraces in the deep narrow valleys, with wheat and barley as the staple crops, but with animals (sheep and goats) in a system of local and longer distance transhumance, and with considerable use of fruit tree crops (apricots, almonds and apples). Economic isolation is associated with political isolation and the very light hand of government in economic affairs, whether from Imperial Delhi before 1947 or from Karachi, then Islamabad, since that date.

In such an area with these difficult environmental and economic circumstances, explorations of the relationships between population and development have traditionally found it appropriate to adopt an essentially Malthusian perspective: there is severe pressure on existing limited resources and there is a fragile equilibrium to be maintained with the environment. Small movements on either side of the population:resources equation can have major consequences. Population growth can easily and quickly move the system from equilibrium into a downward spiral of environmental degradation to threaten sustainability. Indeed it may completely undermine it. In Northern Areas the rate of population growth is currently known to be high: 'While reliable data are unavailable the population of Northern Areas is believed to be growing at three to four per cent per year. Many women bear from four to eight children each' (World Bank 1987:16). The dangers associated with rapid population growth are well recognized in this area, as they are for the whole of Pakistan (Lieberman 1982) and for other Himalayan regions (Blaikie 1985; Stone 1990). However, the activities of development agencies in recent years have been to focus on other aspects of the population/development relationship to promote development, rather than to directly affect quantitative components of population change. In particular they have sought to promote HRD.

The Northern Areas has experienced more substantial economic change in the last 15 years than it probably had in the previous century, and largely as a result of two major and not entirely unrelated developments. In 1978 the Karakoram Highway (KKH) was completed, linking Pakistan with western China by an all-weather road. The road was strategically important to both countries, and built (from internal resources) at great cost to them in the face of enormous engineering obstacles. It has also brought economic benefits to the areas through which it was built, the Upper Indus valley and Northern Areas within Pakistan. It gave to farmers, often for the first time, access to commercial markets outside their local area. The large urban markets of Punjab became accessible to the fruit and vegetables of Northern Areas villages, and the whole area was opened up to trade and the cash economy, including tourism. The two-way flow of goods and cash was inevitably accompanied by flows of people in both directions. It had become easier for migrants to move out of Northern Areas, for, as so widely documented elsewhere, the effect of a road linking a remote and impoverished area with a more dynamic region has tended to be to accelerate if not initiate out-migration, especially of the younger and more highly skilled members of the community (Rhoda 1983). Out-migration from Northern Areas is not new, but its rate has probably been much increased by the KKH.

In December 1982, a major rural development project, the Aga Khan Rural Support Programme (AKRSP), was initiated for Northern Areas (and subsequently extended to include neighbouring Chitral District of North West Frontier Province, an area of very similar environmental and economic conditions). This sought to ensure that local people shared in the benefits the road could bring, by supporting village communities to generate sustainable village economies with rising incomes. The assumptions of the programme were for bottom–up development, focused round broadly-based village organizations (VOs) through which the savings of individual households accumulated and were invested, but further supported by loans and grants to the VOs from AKRSP. AKRSP itself was funded by the Aga Khan Foundation as a non-governmental organization, though it now receives considerable official bilateral donor support, notably from ODA (UK), Dutch, Norwegian and Canadian sources, and also funds from international non-governmental organizations such as OXFAM. Between 1982 and 1988 3456 separate short- and medium-term loans totalling 78m rupees (£2.5m) were allocated to the over 1000 VOs for specific investments in agriculture and marketing (e.g. fertilizer, fruit preserving). In addition, in this same period 717 grants totalling 115m rupees (£4m) have been allocated to productive physical infrastructure projects (PPIs). Over half the PPI grants have been for irrigation and related works, with another 30 per cent for improved roads and tracks. Schools and health facilities are explicitly excluded from eligibility for PPI loans.

AKRSP fits well with current ideologies of rural development. Indeed its establishment preceded much of the development of these current approaches, and the very favourable evaluation of AKRSP by the World Bank (World Bank 1987) may have been influential in reassessing rural development strategies generally. AKRSP is based in the inculcation of essentially market-oriented farming principles. It encourages investment which will lead to individual returns, profits and savings set within a community, small-scale context and with major efforts in environmental management to improve the village systems, the essence of sustainable development.

It is clearly not appropriate here to attempt a full description or analysis of the activities of AKRSP, but rather to note that the general thrust of development initiatives has been to increase the level and range of resources that are to be created and are to be managed, rather than to approach the problem of poverty as a problem of population growth in the first instance. However, within the general objectives of raising productivity, important issues for population have arisen, and need to be addressed by the agencies involved in the development process in the region – i.e. the Government of Pakistan and the Aga Khan network sister agencies of AKRSP, which includes the Aga Khan Education Service (AKES), the Aga Khan Health Service (AKHS) and the Aga Khan Housing Board.

In part due to cultural resistance in such an intensely conservative area but also and more importantly for reasons associated with sensitivity towards and weaknesses of population programmes in Pakistan as a whole, there have been no formal interventions by government or AKHS programmes in fertility or family planning. Attention in the social sector has been directed, implicitly rather than explicitly, to HRD. There are low levels of health and education provision in the modern sector. Religious education has been available to all through mosques of the Sunni and Shia sects, but formal government schooling has until recently been confined to larger towns and villages, attended only by a small proportion of each age cohort, even smaller for girls than for boys. The stock of educated people is low, as are levels of health, as indicated for example by levels of infant mortality that are higher than the already high levels for Pakistan as a whole. Thus the region seems to offer an extreme example of very limited human resource development, at a level that can be expected to affect adversely the potential response to the new opportunities for development that have become available. Strategies for promoting HRD have focused round the two very different approaches identified above: of 'education' through formal schooling and provided by much expanded enrolments in government and AKES schools; and by 'training', particularly associated with the village activities of AKRSP.

In AKRSP there is a very deep-seated and central concern for HRD, explicit in a separate section entitled 'Human resource development' in its quarterly and annual reports, and implicit in its overall objectives. That focus, however, is almost entirely on project-related training. This training takes place mainly in Gilgit, Chitral and Skardu – the three programme offices of AKRSP – and in villages convenient to the trainees, and in training centres of AKRSP, but not in schools. There are courses for members of VOs, chosen by their peers, that offer training to workers in such specific skills as soil management, pesticide use, animal husbandry, irrigation maintenance, etc., all geared to specific PPIs of the village. There is also a women's programme that provides courses in women's activities, such as poultry management. The trainees are explicitly *not* recent school leavers. Young people, who are more likely to have been to school than their elders, are not yet likely to have reached such a status of direct involvement in the VOs, and are thus effectively excluded from the training programmes offered within AKRSP. Conversely the trainees tend to be mature adults immediately responsible for a family or land commitment to the village.

Human resource development is thus operationalized as a practical contribution to directly productive activities, with an immediate feedback from the training to the income-generating activities of the village organization. In its Fifth Annual Review, 1987, AKRSP devoted a major section to 'Human resource development: towards a regional strategy', in which it raised the possibility of collaboration with other Aga Khan network agencies to 'devise a common strategy aimed at skill formation . . . there appear to be opportunities for investing in human capital, so that the human capacity to respond to change may begin to match the challenge of the changing environment' (AKRSP 1988:107). While this statement does identify the need for long-term institution building for skill formation, the first requirement of its suggested programme is that it should 'provide training (at all relevant levels) that is *manifestly useful* to the trainee and the users of his/her services' (AKRSP 1988:108).

Such an intensely practical view of human resource development complements the implicit assumptions of its sister organization, the Aga Khan Education Service (AKES). AKES supports and has recently considerably expanded its commitment to formal education in Northern Areas. It is active both in its own schools and in support for general activities of the government education service in educational planning and teacher training. The strategy of AKES is to foster a broadly-based basic education to as large a proportion of the school-age cohorts as possible up to grade 10, with opportunities for further study elsewhere in Pakistan. AKES has been active in Northern Areas for several decades, and has an honourable record of providing schooling for both Sunni and Ismaili (religious adherents to the Aga Khan) children, and particularly for girls in a situation where government resources have been very strongly biased towards boys' schools. Its potentially long-run and essentially infrastructural approach to human resource development continues to be strongly attractive to the local population and enrolments in the AKES schools continue to grow. It is certainly the case that in Pakistan, even more than in many Third World countries, formal education is very 'academic' and examination oriented. The 'diploma disease' has a strong hold on parents and school children, and there is very little practical or vocational education in the curriculum. Quality of achievement is low, and there is general ill-ease about the national implications of a system that fails to cater for the majority of the nation's children and seems to yield low economic returns to the individual and to Pakistan as a whole. Nevertheless, there seems to be at the national level a positive relationship, even though rather weak, between levels of farmers' education and farm income (Butt 1984). Nationally the formal system generates an apparent over-supply of school leavers in the formal job market, though the problem of educated unemployment is not yet apparent in Northern Areas. The government does have training schemes and business promotion grants targeted at school leavers, e.g. the Youth Investment Promotion Scheme (YIPS), but depends mostly on on-the-job training for industrial and commercial skills.

In Northern Areas AKES has made serious efforts to improve the educational experience in its own schools and in government schools. Innovation have included improvements within the existing curriculum (e.g. in raising the quality of the teaching through in-service teacher training) and also through curriculum change that seeks to link the school more directly with the local economy and society. Given the particular and serious problems of human use of the environment of the region, one area of current interest is in environmental education, to provide a more systematic knowledge of the nature of the problems and how they might be managed. Here, more than almost anywhere, there is potential for application of the sort of curriculum advocated by the Brundtland Commission on Environment and Development:

> Rural schools must teach about local soils, water and conservation of both, about deforestation and how the community and the individual can reverse it. . . . Education should therefore provide comprehensive knowledge, encompassing and cutting across the social and natural sciences and the humanities, thus providing insights on the interaction between natural and human resources, between development and environment.
>
> (Brundtland Commission 1987:113)

Here is the basis for a curriculum in Northern Areas that is of general rather than of specific relevance, and needs to be seen in conjunction with the more direct and complementary training activities within AKRSP. It is similar to more extensive programmes in environmental education developed in other Third World countries where there are similar critical population/environment issues to be tackled, such as in Ethiopia (Fitzgerald 1990).

It is in the area of migration that the differences between the 'education' and 'training' strategies are likely to have greatest effect. The 'training' strategy assumes low levels of out-migration, that the trainees remain in or will return to their villages, and are selected to ensure that. While the aim of curriculum innovation in the schools is to promote a similar outcome, to ensure a closer relationship between the school and the needs of the local community, there is little doubt that increased formal schooling will have the effect of increasing the rate of out-migration of school leavers, for it seems to provide those who have been to school with skills (i.e. literacy and numeracy) and attitudes that are more appropriate to urban than to rural areas. For those with some education, migration becomes a more attractive option. Formal schooling, especially beyond primary school gives school leavers and their families job expectations and personal aspirations that cannot normally be met within the region and require out-migration to the towns and commercial farming areas elsewhere in the country. During its long period of isolation there were relatively low rates of migration out of the region, but with impulses associated with the Karakoram Highway and the commercialization of agriculture in the last decade, it is probable that the rates of out-migration and also in-migration have sharply risen. Out-migration has been accelerated by the educational expansion of recent years, such that school leavers are probably a growing, though still small, proportion of the total flow. It is known that some school leavers do leave the region to find further educational opportunities elsewhere, as well as jobs directly. Spatially, it seems that the migration remains, as it has been for many decades, strongly biased to Karachi as the principal destination. This is not the nearest major town, but it has strong links with Northern Areas through the Ismaili community and personal and family contacts. These links are particularly valuable for those who migrate to Karachi for training as engineers, accountants, nurses, etc., after *matric* at grade 10 (Selier 1988).

While this may be construed as rural brain drain, a net loss of educated young people from the region in the short term, and would certainly be so if the moves were to be permanent, it takes too narrow and short-term a view of the migration process, for there is the possibility of return. The possibility is greatly enhanced by the development impulses outlined above, for a commercialization of the rural economy has directly generated a demand for trained and experienced accountants, engineers and other technicians, as well as for teachers, nurses and other personnel for whom training opportunities and the ability to acquire appropriate experience in Northern Areas are distinctly limited. At present there is a local shortage of skilled personnel, with some posts filled by people from 'down-country', so that the opportunities for returnees with skills are both considerable and growing. There is currently a very major problem of lack of data on the nature of the local labour market and of the migration system as it operates in the region. Research is needed to examine the parameters of the incidence and nature of the initial migration and subsequent return of school leavers, and this can have major value for enhancing the demographic knowledge base on which planning decisions can be based. Under what circumstances are local school leavers returning to Northern Areas after further training and experience elsewhere?

'Education' and 'training' are clearly both necessary for human resource development in Northern Areas. Education is targeted to a much wider population in the schools, with generalized benefits for agriculture and other economic activities. Training is targeted at the specific needs of those directly active in the villages of the region, and is necessary for the success of AKRSP projects. A school-based education, however, will be more likely to further a broader range of social and economic objectives (including improving health and reducing fertility) and raising farmers' awareness and the actual introduction of new technologies and the potentials for rising incomes. Human resource development must continue to promote formal schooling as an infrastructural investment for the long-term sustainability of the region.

Western Province, Kenya

Both Kenya and Pakistan are amongst the 41 countries in the poorest category in the World Bank classification of countries, Pakistan being 24th poorest with an estimated per capita GNP of $370, and Kenya the 23rd poorest with a per capita GNP of $360 in 1989 (World Bank 1991a:204) The general relationships between population and development have been given much greater prominence in Kenya than in Pakistan, both by government (Kenya Government 1984) and by population analysts (Frank and McNicoll 1987). Kenya has experienced extremely high rates of population growth in recent years, generally in excess of 3.5 per cent during the 1970s and 1980s, and fed particularly by what was by 1979 a total fertility rate of 7.9, the highest of any country included in the World Fertility Survey (WFS). The TFR had by 1989 fallen substantially to 6.7, as measured by the National Demographic and Health Survey (DHS) (Kenya Government 1989a), but in global terms remains extremely high. The Government and most commentators have adopted a distinctly neo-Malthusian view of the Kenyan demographic situation, promoting fertility reduction as a major thrust of population policy, and bringing that population policy close to the core objectives of overall development strategy (Kenya Government 1989b). Population policies in Kenya are not strong, at least in comparison with the vigour to which they have been pursued in several East Asian countries, notably China, but they have been more forceful than elsewhere in Africa (World Bank 1986). That the fertility rate has fallen 13 per cent between 1979 and 1989 may be due in part to direct population policies, in particular to rising contraceptive prevalence rates, but also to changing patterns of demand and family/household relationships, raising controversies about the bases of fertility behaviour and the possibilities for further fertility change that are now of major urgency for research work in the country (Kelley and Nobbe 1990).

Some of these basic demographic trends in Kenya were identified in Chapter 5, but what is important in the context of the present discussion is that the apparent fall in fertility rates has been strongly differentiated by province. Reductions have been concentrated in Nairobi, the capital city, and in rural provinces of greatest commercial development, Central and Rift Valley. While the national TFR fell from 7.9 in 1979 to 6.7 in 1989, it fell from 8.2 to 6.0 in Central but remained at 8.1 in Western Province for both survey years (Kenya Government 1989a:24). Western Province is the poorest area of Kenya to be included in the DHS. It is a province with a population of over 1 million at very high overall densities of over 250 per km^2, rising to over 750 per km^2 in parts of

Kakamega District (Fig. 6.3). Here is an area with relatively benign environmental conditions (in sharp contrast to Northern Pakistan), with well-distributed and regular rainfall and well-developed volcanic soils. These support a subsistence agriculture based on maize as the staple food crop, but with a range of subsidiary foods. Recently there has been some development of cash cropping, notably of tea and sugar. Nevertheless there is

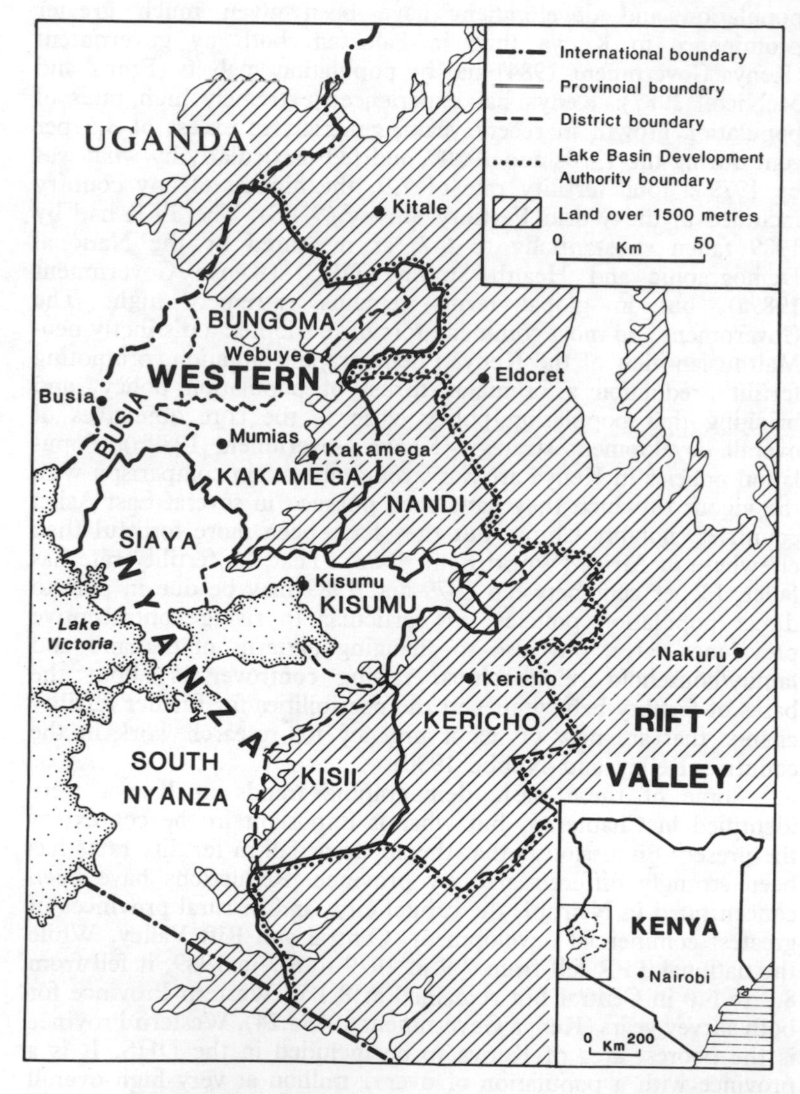

Fig. 6.3 Western Kenya

substantial population pressure, particularly in the highest density areas of southern Kakamega (Ominde 1972), such that a falling rate of population growth might be expected to be a rational response to that pressure. However, this has not been the preferred response of the local population, for natural rates of growth continue to be high, without apparent major disequilibrium of the local production system. While there are many possible reasons for the lack of demographic response (whether due to weakness of supply factors, i.e. a weak implementation of family planning programmes, or to lack of demand) that require detailed investigation, one major feature of population change in the province is the very considerable importance of out-migration. This is overwhelmingly circular migration and mostly to urban destinations, for family survival in much of the province is heavily dependent on remitted income (Gould 1988a, 1992). Since migration is the principal means of maintaining the household economy for most families, smaller families will mean fewer migrants and lower household incomes. Most families have pursued a human resources response rather than a demographic response to the problems imposed by poverty and land shortage.

Human resource development in Kenya is strongly associated with employment policy. Chapter 9 of the current National Development Plan is entitled 'Employment and human resource development' (Kenya Government 1989b:193–224). An economy of growing complexity and technological sophistication needs a more educated and skilled population, and the emphasis in HRD is much more towards education and training than to improving health status, though that is certainly not ignored. There have been very substantial expansions in enrolments in education at all levels since independence in 1963, both in absolute and relative terms, such that enrolment expansion has more than matched the growth of the school-age cohorts (Gould 1986; Uitto 1989). Public and private expenditure on schooling are very high (over 30 per cent of national government recurrent expenditure was devoted to education in the 1980s, but that proportion is now falling). It is largely allocated to a formal system with a traditional 'academic' curriculum. Recently 'agriculture' has been introduced as a compulsory subject in the primary school curriculum, but the system remains strongly biased towards creating a base of general education rather than a formation of specific skills.

Although the vast majority of schools at secondary level follow an obviously traditional academic curriculum, there are some well-funded and equipped secondary technical schools. However, the graduates of these schools have been shown to have no greater propensity to find employment than graduates of the standard secondary schools. Furthermore, their level of skill acquisition as a result of their being in technical secondary school for four years is not greatly enhanced (Närman 1988). In Kenya, as elsewhere, the substantial additional cost of providing a diversified and 'relevant' secondary curriculum is called into question. At the tertiary level, however, manpower planning considerations are critical in determining the range of training available. Yet even here, as Bennell (1986) has demonstrated, the formal provision in engineering produced graduate engineers but the real need was for lower level technicians in the industrial sector. The national policy has been premissed on the neo-classical infrastructural conceptualization of a broad-based education for as large a proportion of children as possible.

More generally the expansion in secondary enrolments of whatever type, in a country where the growth in formal sector employment has not been spectacular and certainly less than the output of the schools system, has inevitably produced an escalation of job qualifications (Hughes 1991). Large-scale surveys of the labour force in Nairobi and Dar es Salaam in Tanzania in the early 1980s record substantial escalation in both capital cities, but more in Nairobi. Yet there is still a strong positive relationship between education and earnings. In every occupational group the educated are still paid substantially more than the less educated. Hazelwood concludes:

> The fact that they are paid more in the same occupation suggests that they are more productive. If this is so, the filtering down of the more educated raises the productivity of the occupational group they enter. Their additional education is not wasted if they filter down.
>
> On-the-job training cannot, it seems, be relied upon to substitute for the beneficial effects of education on productivity.
>
> (Hazelwood *et al.* 1989:282)

More education is clearly beneficial and 'rational' for the individual.

Western Province has shared fully in this thirst for formal education. Though relatively poor in income terms, it achieves above the national average in proportions of children enrolled at all levels (see Fig. 3.3). Over 80 per cent of boys and girls are enrolled in primary schools. There is a massive demand for schooling that government has been unable to satisfy and as a result there has been a major development of self-help, *harambee* schools that have required allocation of even more family and community resources. In many areas of the province over 20 per cent of household income is allocated to educational expenditure for, in the absence of income-earning opportunities from extremely small family plots in high density areas, education is considered to be a much more productive investment than agriculture. It is the educated who are more likely to migrate to seek and to find a job elsewhere in Kenya, and to remit part of their earnings to the home area (Gould 1985a), and in such a situation educational expenditure can be seen as consumption, for short-term survival, as well as investment (Martin 1982). Investment by families and households in HRD requires that the benefits are earned from employment elsewhere, but for the long-term benefit of Western Province. Migration is the prime mechanism through which the human resource base can become productive.

This external view of the purpose of HRD is not new for Western Province. Migration of wage labourers has been very much part of its economic life since the colonial economy was first developed at the beginning of the present century. In the colonial period the education system was largely controlled by Christian missions who sought, with colonial government support, to develop technical and vocational education that would give trainees skills that would be useful in rural areas. These were not only in agriculture directly but in building trades and other services. However, their trainees, too, became migrants, for the skills of carpenters, masons or tailors were in even greater demand in towns than in these impoverished rural areas. Thus the more practical, training-based approach to education also became a training for migration, quite the opposite of its original objectives (Gould 1989).

Since independence the bias shown by the population at large against training for locally relevant occupations has persisted. As was discussed in Chapter 4, in Kenya as a whole from the 1970s there was a rapid growth of village or youth polytechnics, the principal aim of which was to provide primary school leavers with skills that they would apply to enhance their economic prospects in rural areas. Many village polytechnics were established in Western Province in the 1970s, but they too were seen as second best compared to the formal, academic schooling, and not a 'shadow' system of equivalent status. Polytechnic trainees seek skills that are thought to equip them to find jobs in urban areas, mostly outside Western Province (Gould 1989). This may in practice prove to be a training for unemployment, given the excess of school leavers and trainees over available jobs, but investments in formal education and polytechnic training continue at a substantial level in the province, as throughout Kenya. Foster's vocational school fallacy finds a strong echo in Western Kenya. The education system is used not only to enhance the human resource base of the province, but also to ensure its integration with the national labour force.

In Western Kenya there clearly is a strong link between HRD, whether based on formal schooling or specific training, and migration. There is, however, a very weak link between education and the impoverished and marginalized rural economy. National development strategies have sought to integrate the province more fully into the national economy and this has been supported from above by government – in development of a well-structured urban and commercial hierarchy – and from below by families promoting migration through investment in education. HRD is strongly directed to the formal urban employment sector.

Policies for human resource development

The policies for HRD and education have rather different emphases and strengths in Northern Pakistan and Western Kenya. Both regions are poor and peripheral to the main areas of their national economy, but there are also major differences, notably that Northern Pakistan has much less favourable environmental conditions for endogenous development and is much less well integrated into the national economy than is Western Kenya. Its human resource base has been less well developed, with much less education for far fewer people, both boys and girls, but particularly for girls, and modern health provision is also less available. Nevertheless there is much more explicit concern in Northern Pakistan for HRD to support local economic impulses, both in the specific training courses offered for village workers and in recent curriculum proposals geared to better and more relevant environmental education in schools. This is expected to reduce out-migration, and eventually to encourage some return migration of previous school leavers who had left the area and had acquired further qualifications, experience and capital elsewhere. The trainees can then take advantage of the new opportunities available as a result of the increased commercialization of rural production. HRD in both education and training is part of a wider strategy of promoting self-sustaining and sustainable local development. It has to be seen in the context of a project-related response to broad rural development strategies, led by the activities of AKRSP. In Western Kenya, by contrast, HRD has an essentially external bias, to support the migration system, by providing in schools rather than in specific skill training the basic literacy and numeracy and higher 'academic' qualifications that urban migrants require in order to begin their job search. Its presumed direct relevance to local rural development is limited, even in the longer term.

In both these cases and for the countries of the Third World taken as a whole, there has been weak integration between HRD and broader education policies. They seem to be pursued largely independent of each other. Policies for HRD are concerned primarily with population quality and how it can be enhanced to assist the national development effort. Policies in education have in the past been concerned primarily with the size and shape of the schools system and how numbers and distributions of enrolments can be managed. HRD policies are premissed on an essentially optimistic and positive view of population as a resource that can be better used for development. By contrast, education policies are generally felt to be based on a rather negative view of population, as a financial burden in the short term, sufficiently large that the long-term advantages of the raising of skill levels may not be realized. These apparent incompatibilities can and should be resolved, otherwise they will prevent their effective integration into an overall policy that encompasses and weighs both the costs and benefits of overall changes in numbers and skills of the population. These conflicts confirm the necessity for the approach to human resource development to be broadened. It needs to move from the narrow confines of the formal system and a quantitative approach that has focused almost entirely on meeting the social and political imperatives of school and training provision. It needs to move towards a more qualitative approach that considers actual and potential production of the educated and trained. The impact of HRD policies varies substantially within and between regions as a result of different local development opportunities. A broader approach will therefore of necessity need to be more sensitive to local variations in population/development relationships than traditional approaches have been (Farooq and MacKellar 1990).

Policies in education are central to the management of the human resource. They will affect not only the extent of the national resource enhancement through the size of the system and the numbers of people exposed to any learning experience, but they will also affect the nature and type of that experience. Improving the quality of education and training programmes to ensure rising levels of real achievement, learning and skill acquisition to more flexibly meet the needs of a rapidly changing economic system must be a major priority. Human resource development must be more about raising levels of competence than about raising levels of qualification.

Chapter 7

Education and migration

Education adds to the value of the human resource base by raising the skill levels of those who have attended schools or have been involved in some specific formal or informal training. The enhanced population resource is not of great economic value in itself, and it needs to be linked to other productive resources to take advantage of the newly acquired potential, and therefore to allow its new value to be realized. The bringing together of different resources is an essential feature of any production process, and requires that at least some of these resources are mobile. People are the most mobile of productive resources, and it is this mobility that gives enormous importance to the human resource base in the development process. There is a strong and positive link between education and mobility: other things being equal, the educated in any population are more likely than the less educated to become migrants. This is borne out in a mass of empirical evidence in developed countries, e.g. in the United States (Shryock and Nam 1965) as in the Third World (Connell *et al*. 1976; Simmons *et al*. 1977). It is to this strong link that this chapter now turns, to examine how this relationship operates in the Third World and with what effects on development at the national and international scales.

The previous discussion of human resource development showed how education and training might have an immediate effect on development in the areas where the educated live and receive their education. This applies particularly to agriculture and rural development generally, and schemes designed to offer skills to young people in rural areas that can be used within their own communities are familiar throughout the Third World. Yet it is apparent from the discussion in Chapter 6 that many of the people involved in rural education and training acquire skills or a general background that provide them with access to a wider range of better paid and generally more attractive opportunities elsewhere. Their skills are spatially as well as occupationally transferable, and educated people move out of the communities in which they acquire the skills, even though the skill training may have been initially designed to prevent migration. How much more obviously this applies to skills that are associated with urban and modern sector development. Formal schooling provides, as a minimum, formal literacy, especially in a national language where it is not the local *lingua franca*, and basic numeracy, and the longer a person remains in the school system the more 'modern' the skills become. The demand for people with these higher, scarcer and more 'modern' skills is greatest in areas of modern commercial urban and rural development, and these are the areas that attract skilled people.

Just as 'education' is not an unambiguous term, neither is 'migration'. Population mobility can comprise a wide variety of phenomena, differentiated by directions of movement (a four-fold mix of rural and urban sources and destinations) whether it is temporary (circulation) or permanent (migration) within any period of time, and whether it involves moves within countries or between them. These basic dimensions are identified in Table 7.1. Education-related moves can be those which occur in the process of acquiring an education (and have been discussed in Chapters 3 and 4), and comprise circular, non-permanent moves between home and school, most commonly daily and, for the largest number of people rural–rural, from rural homes to nearby rural schools. With higher levels of education, when schools and colleges tend to be located in central places, the moves are more likely to be rural–urban or even urban–urban in the case of higher education in

Table 7.1 Typology of population mobility: education-related movements

	Internal		International	
	Circulation	Migration	Circulation	Migration
Rural–rural	Primary school children			
Rural–urban	Secondary school children	Educated job seekers		
Urban–rural	National service workers	Retirees		Retirees
Urban–urban	Intra-company transfers	Inter-company transfers	Specialist trainees	Skilled technicians

institutions. Decisions about the periodicity of the move, whether it is daily, weekly, termly or even longer is in part a function of distance between the home and the institution in question (the less common the institution, the greater the likelihood of longer distance moves and longer periods of temporary absence). There are, in addition, cultural factors associated with the traditional roles of children in the household or attitudes to child fostering with relatives that may affect the form of circulation, e.g. in Peru (Skeldon 1985), in Ghana (Brydon 1985), in the Mahgreb countries (Sutton 1989), in East Africa (Gould 1974, 1985b). These circular moves for education are almost entirely internal, but some may be international to specialist training available only outside the country, and will tend to be from urban area in the source country to an institution in an urban area in the receiving country. It will involve circulation where there is a presumption of return home at the end of the course abroad.

This chapter, however, is primarily concerned with those moves of the educated in search of employment after they have completed their education or a period of training. Here too there may be a presumption of return to the home area after a period of work away, as was discussed in Chapter 6 in the case of school leavers in Northern Pakistan. In that case the moves were almost entirely to urban destinations (predominantly Karachi), and the majority of migrant school leavers return to their rural origins to live after a period of absence for further study or employment/unemployment. Although there may be opportunities in rural areas for educated job seekers, most of the migration will be rural–urban or, increasingly with a larger proportion living in urban areas, urban–urban. Movement between urban centres is often hierarchical, from smaller to larger centres in 'step-wise' movement. Government postings and intra-company transfers of skilled managerial or technical staff can move individuals temporarily up and down the urban hierarchy to postings in different towns at various career points. The larger the country, the more complex the postings and transfers are likely to be. In several Third World countries (e.g. Nigeria, Nicaragua, Indonesia) there are national service schemes, variously named but similar in purpose, that promote an urban–rural temporary movement of newly appointed or potential government employees such as teachers, doctors or nurses, to undertake a period of service in rural schools or health centres, respectively, before posting to an urban school or hospital. These, however, are very much the exceptions rather than the rule, for there is an overwhelming rural–urban flow of the educated in search of the job opportunities available in urban areas for those with the relatively scarce skills.

At the international scale, movements of the highly educated are also associated with intra-company transfers of staff within multinational companies, mostly on temporary assignment, but most attention in this category must be focused on the 'brain drain' – permanent movements of highly skilled professionals in an international labour market from poor to rich countries. Almost inevitably these are confined to urban sources and destinations. The 'brain drain' is discussed in the concluding section of the chapter.

Internal migration: selectivity by education

The problems of separately identifying the effects of education on population behaviour has been a recurring theme of earlier chapters, and they are as difficult in the case of migration. The problem of multicollinearity is as familiar in migration analysis as in educational studies generally, and has been approached through the use of a multivariate format, though not necesarily a formal statistical model, to measure the contribution of education as one of a range of independent variables to 'explain' migration, measured as a dependent variable. The analytical problem of multicollinearity with migration has been well illustrated in a comparison of migrants within, out of and into four highland communities in Ecuador. The educated are the most prominent group in all three migration categories in all four communities, but the differentiation of migrants from non-migrants is also related to a wider range of social end economic variables:

> . . . in every community studied, those with most education came from families with the most economic resources: they owned most land or had alternative sources of superior incomes. In every community, those with most education lived in the village centre, or very near it, while those with least education lived in remote hamlets or isolated farmsteads. Those with most education came from families who themselves had high educational levels and were likely to have had migratory experience themselves. Those with most education were young, *mestizo* and male.
>
> (Preston 1987:204)

Migration studies at this scale in the Third World have emphasized the importance of 'economic' factors as a principal cause of movement, with 'education' measured in number of years in school or qualification achieved, as a proxy for earning capacity. This variable is typically set alongside income levels, employment totals and urbanization as the critical 'economic' factors at the aggregate scale (Gould 1992; Todaro 1976). In the many country-specific spatial interaction models of migration, education was generally shown to be of fairly small importance. In Uganda, for example, using 1969 Census data to explain migration as measured by a 15 × 15 place-of-birth by place-of-residence matrix, a gravity interaction model, with distance, population size, urbanization proportion and income level as independent variables, accounted for 73 per cent of the variance in the migration data. When an education variable was added to the model, with levels that were strongly correlated with income and urbanization, the explantation rose only to 75 per cent (Masser and Gould 1975:84). Despite these and other disappointing results, aggregate model builders have usually sought to build an education variable into macro-economic formulations of migration/development relationships (Moreland 1984).

An alternative approach at the aggregate scale is to reject the apparently unidirectional multiple regression format, with migration being 'explained' at least partially by educational status of the population, in favour of more open systems conceptualization with migration affecting as well as being affected by education, as in Fig. 7.1. Expansion of urban job opportunities and a growing rural–urban gap in wage levels generates demand for more rural schools, centrally by government and locally by the communities themselves. A spontaneous adjustment might be expected in the medium term with the urban demand slowing and the rural–urban income gap narrowing for educated workers where there was large and rapid expansion of school places, thus creating conditions for excessive urban migration of the educated, and rising unemployment. Migration in these terms is clearly both caused by and itself affects levels and patterns of school enrolments. The decisions to educate and to migrate are not at all independent of each other.

There are clearly problems in attempting to deal with the migration/education relationship at the aggregate scale, but these are partially resolved with the adoption of a behavioural perspective that considers individual migration decisions and the role of various factors, including education, in these decisions (Brown 1991; Gould 1992). Both migration and education need to be examined in the context of continuous and continuing economic, social and demographic change, and in this context against a background of changes in the quality and type of education migrants receive. Also relevant are changes in the out-of-school environment in which they find themselves once they leave school. Within the schools teacher quality, curriculum change, greater availability of textbooks and other teaching materials will all, separately and together, affect the people who have been to school; outside the schools, the demand for educated people in urban areas is apparently falling in most countries in the face of sluggish economic performance and a changing occupational structure. Data on individual migration behaviour may be collected either in cross-sectional or longitudinal

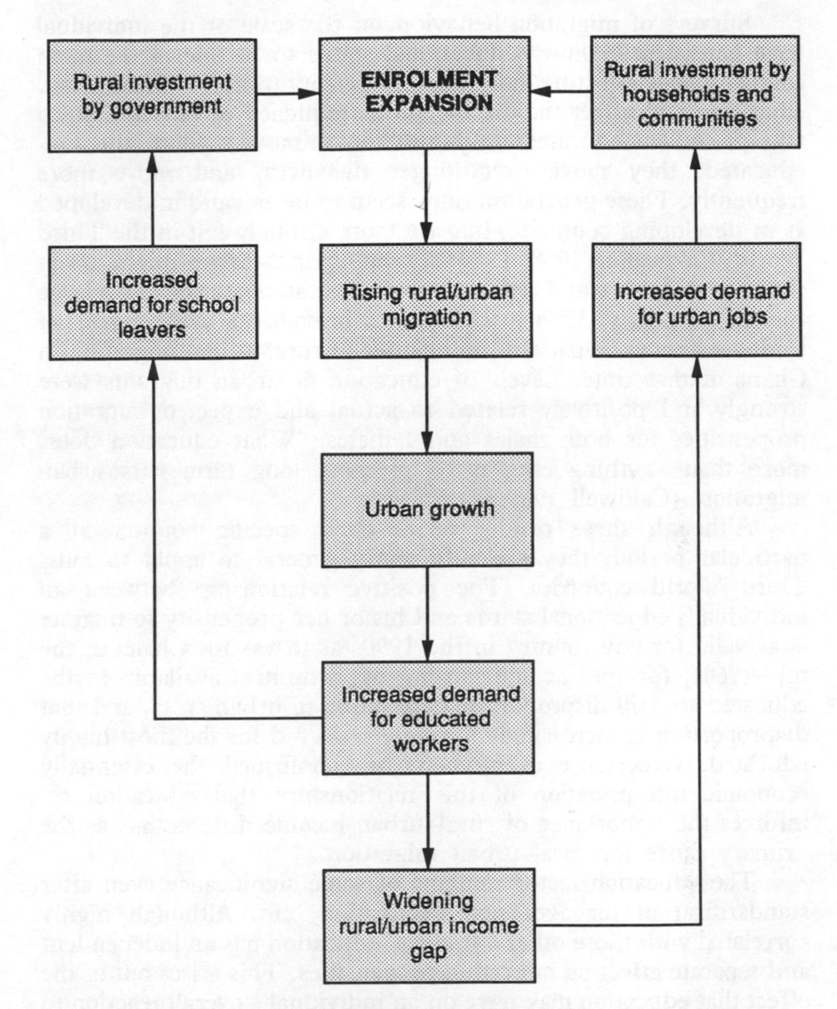

Fig. 7.1 Education and rural–urban migration: some relationships

studies. Cross-sectional studies of a general or purposive (i.e. educated) sample are the most common approach and have provided most of what we know of the mechanisms of the education/migration relationship. However, given these changes within and outside the school, a longitudinal approach that places the change over time at the heart of the methodology needs also to be considered. In this respect 'tracer studies' assume a particular importance, not only as a technique for manpower planning but also as a means of investigating the education/migration relationship directly.

Surveys of migration behaviour at the scale of the individual have consistently shown educational status to be one of the main characteristics affecting individual propensity to move. In addition, education will affect the distance and frequency of moves that do take place. The educated are more likely to move than are the non-educated; they move over longer distances, and move more frequently. These general findings seem to be as valid in developed as in developing countries, but are more strongly felt in the Third World. Caldwell's (1968, 1969) survey of rural–urban migration in Ghana, 1962–64, was a classic national migration survey. This large survey of nearly 17 000 individuals provides a wide range of evidence about the patterns and causes of rural–urban migration in Ghana at that time. Levels of education of urban migrants were strongly and positively related to actual and expected migration propensities for both males and females. 'What education does, more than anything else, is to promote long term rural/urban migration' (Caldwell 1969:62).

Although these results relate to a specific country at a particular period, they are sufficiently general to apply to most Third World countries. The positive relationship between an individual's educational status and his or her propensity to migrate is as valid for any country in the 1990s as it was for Ghana in the mid-1960s, for the income-earning opportunities available to the educated are still disproportionately found in urban areas, and that disproportion is increasingly strongly observed for the most highly educated. Experience everywhere has confirmed the essentially economic interpretation of this relationship: that education re-inforces the importance of rural–urban income differentials as the primary cause for rural–urban migration.

The education factor remains of some significance even after standardization for age, sex, occupation, etc. Although highly correlated with those other variables, education has an independent and separate effect on migration propensities. This arises out of the effect that education may have on an individual's overall reaction to real and perceived opportunities. The search space of the educated migrant would be expected to be widened as a result of his or her having been to school, whether as a result of leaving home to attend school or as a result of exposure to new and potentially socially unsettling influences. For example, the role of education in the development of a perspective of individualism rather than the communalism will be reflected in attitudes of educated people to the type and purpose of remittances and investment of savings in rural and urban areas. In Botswana, for example, the urban economy has been much more buoyant than the rural economy, and the educated have been most aware of urban commercial opportunities (Bell 1980). Given better returns to investment in rural areas, the educated would be as willing as the élite were in traditional Tswana Society to make rural investments. They have also done so in the commercially vibrant small farm economy in Central Province, Kenya (Collier and Lal, 1986). There is no mass rejection of the rural economy for other than economic reasons.

The relationship between education and migration, though normally positive, is not linear. Many studies have identified a 'J' shaped curve: those with education or only a few years in school have higher migration propensities and migrate over longer distances than those with only higher levels of primary education (Fig. 7.2). There are important points of rapidly increased migration at the end of school cycles, hence the large numbers of studies that focus on migrations of particular groups of leavers: primary school leavers, middle school leavers, or secondary school leavers (Gould 1982b).

The educational characteristics of rural–urban migrants can be expected to be different from those of rural–rural migrants from the same area. Caldwell's survey was specifically to investigate rural–urban migration, as have been many of the other studies cited in this chapter, yet migration within rural areas probably involves more people in all countries. In some countries, e.g. Indonesia, rural migrations may have a greater overall economic impact than

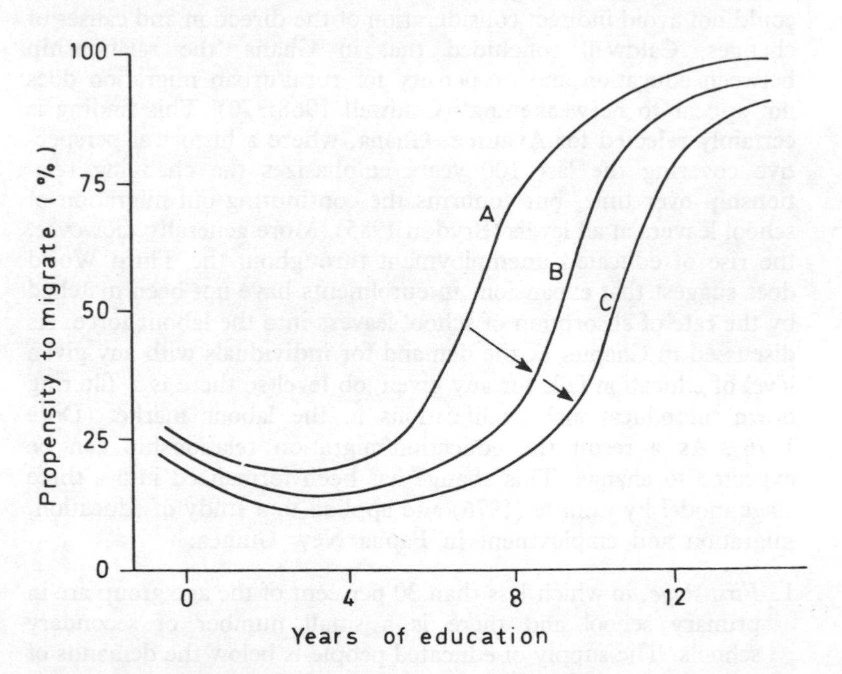

Fig. 7.2 Propensity to migrate by years of education

do rural–urban migrations. Since rural migrants have traditionally been assumed to be unskilled rural labourers or settlers in new settlement schemes, their educational status has been taken to be largely irrelevant, and therefore ignored. Lipton suggests that 'educated to the big city, illiterate to rural areas' is a valid summary (Lipton 1980:5). However Rosemary Preston (1987) showed that in Ecuador some rural destinations attracted the better educated, a feature ascribed to ethnic rivalries. The educational selectivity of migration can vary considerably from region to region or village to village in any given country. Caldwell (1969:49–50, 63–7) identifies the extent of contrasts in education/migration relationships for four major divisions of Ghana. Preston (1987) similarly illustrates differences in Ecuador at the village scale. In both cases the variation is related to overall levels of educational and other socio-economic development in each region or village and to local opportunities for the educated. Thus the national migration propensity/education curve is shown to be merely the average of a wide range of geographically differentiated curves.

The educational selectivity of migration tends to vary over time for any one country or area. Since most large-scale migration surveys have been cross-sectional they have not sought directly to consider changes in migration behaviour over time, though they could not avoid indirect consideration of the direction and causes of changes. Caldwell concluded that in Ghana 'the relationship between education and propensity for rural/urban migration does not appear to be weakening' (Caldwell 1968:370). This finding is certainly rejected for Avatime, Ghana, where a historical perspective covering the last 100 years emphasizes the changing relationship over time, but confirms the continuing out-migration of school leavers at all levels (Brydon 1985). More generally, however, the rise of educated unemployment throughout the Third World does suggest that expansions in enrolments have not been matched by the rate of absorption of school leavers into the labour force. As discussed in Chapter 6, the demand for individuals with any given level of education falls for any given job level so there is a 'filtering down' of educational qualifications in the labour market (Dore 1976). As a result the education/migration relationship can be expected to change. This change has been formalized into a three stage model by Conroy (1976) and applied to a study of education, migration and employment in Papua New Guinea:

1. *First stage*, in which less than 30 per cent of the age group are in primary school and there is a small number of secondary schools. The supply of educated people is below the demands of industry, commerce and government for their services. In such a

situation even primary school leavers have a high chance of finding a job in urban areas, so that there is a steeply rising propensity to migrate with education, and the steep rise begins at low levels of educational attainment.

2. *Second stage*, when 50–80 per cent of the primary school age group is enrolled and there has been a disproportionately large expansion of secondary school places. Since the numbers of school leavers coming on to the job market rises considerably, urban unemployment appears for the first time. There will be falling migration propensities for the uneducated and those with low levels of education. At this stage expansion of higher level job opportunities with the replacing of expatriates by local citizens will maintain the need for higher secondary and university graduates to match demand for expansion of enrolments at these levels.
3. *Third stage*, 'The third stage commences when universal primary education has been experienced for long enough to permit a substantial proportion of the population to undergo schooling. By this time it is hoped that the belief that primary education is merely a necessary condition for involvement in certain monetary sector activities (rather than a sufficient condition for lucrative wage-employment) will have become widespread' (Conroy 1976:23). The expected result will be a further fall in migration rates for primary school leavers and correspondingly for secondary school leavers.

The trends implied by this model are identified by the three curves in Fig. 7.2. The basic 'J' shaped curve of the migration propensity/education relationship moves over time not only downwards but also to the right. Migration propensities fall marginally for the uneducated, whose propensities are low throughout, and for university graduates, whose propensities are high throughout, as there is even in stage three a continuing shortage of highly skilled people. The sharpest movements are in the middle range, affecting those with 6–10 years of education, the categories most affected by the filtering down of educational qualifications as a result of rapid expansion of enrolments.

The development of the third stage of the Conroy model remains a matter of some controversy, for there is a wide range of experience. In the Niger Delta area of Southern Nigeria for example, an area of overall population decline since 1960 but with near universal primary education over the last 30 years, primary school leavers continue to migrate as they would be expected to in the earliest stages of the model, many finding opportunities in the vibrant informal urban economy in Southern Nigeria (Chiegwe

1985). In Ghana, too, there is still migration of primary school leavers into towns, in this case, as in Nigeria, into informal apprenticeships (Brydon 1985).

Change over time in migration propensities is most appropriately identified in tracer studies. These identify and compare the progress of successive cohorts of school leavers into and through the labour market. One such cohort study of primary school leavers examines an impoverished, high density area of Western Kenya, but where the majority of children (boys and girls) attend and usually complete the primary school cycle. Using school registers as a source, a sample of primary school leavers from eight schools in 1971, 1973, 1975, 1977, 1978, 1979 and 1980 was identified in 1982, and their migration and occupational histories in the period between leaving primary school and the period of the survey reconstructed (Gould 1985a). This yielded a great deal of information about their migration patterns (predominantly to Nairobi despite the growing availability of jobs in nearer towns), their periodicity (still, as in the colonial period, predominantly temporary circulation rather than permanent migration), and their causes (mainly to seek further education opportunities in the years immediately after leaving primary school, but dominated by job-related mobility after four years). In the context of the present discussion of changes over time, however, comparison of the eight cohorts suggests a continuity, and if anything an increase rather than a decrease in mobility of primary shool leavers over time (Table 7.2). Even allowing for problems of recall and memory lapse for the earlier cohorts, the general trend of moves per person is upwards for all moves (i.e. including educated-related moves), but there is no obvious trend for work-related moves only.

More direct comparisons of the migrations of the cohorts are possible where the data are ordered, as in Fig. 7.3, by the number of years since leaving school. In this diagram the place of residence of the sampled individual in mid-year is recorded in five spatial categories, to include Tiriki, the home area, elsewhere in Western Province, in Rift Valley Province, in Nairobi and elsewhere in Kenya. The continuing importance of Nairobi as the principal destination is clear from the place of residence of members of the 1971 cohort more than four years after leaving primary school. In all cases Nairobi is the principal destination, and seven years after leaving there are more individuals in Nairobi than in the home area. A similar pattern is evident for the 1973 cohort. The first few years for the later cohorts suggests that the progress of the migration is taking a similar course to that of the earlier cohorts.

The overall pattern of migration destination preferences of a broader sample of pupils in the final year of primary school in Tiriki at that time is illustrated in Fig. 7.4(a). This can be

Table 7.2 Moves per person and per year by cohort of primary school leavers: Tiriki Location, Kenya, 1971–82

Cohort	All moves			Job-related moves		
	Total	Moves per person	Moves per person per year	Total	Moves per person	Moves per person per year
1971 (n = 12)	46	3.83	0.35	27	2.25	0.23
1973 (n = 9)	39	4.33	0.38	15	1.77	0.20
1975 (n = 10)	44	4.40	0.63	24	2.40	0.34
1977 (n = 12)	34	2.38	0.57	13	1.08	0.22
1978 (n = 8)	17	2.13	0.53	13	1.63	0.20
1979 (n = 11)	16	1.45	0.48	5	0.45	0.15
1980 (n = 13)	16	1.25	0.62	8	0.62	0.31

Source: Gould 1985a.

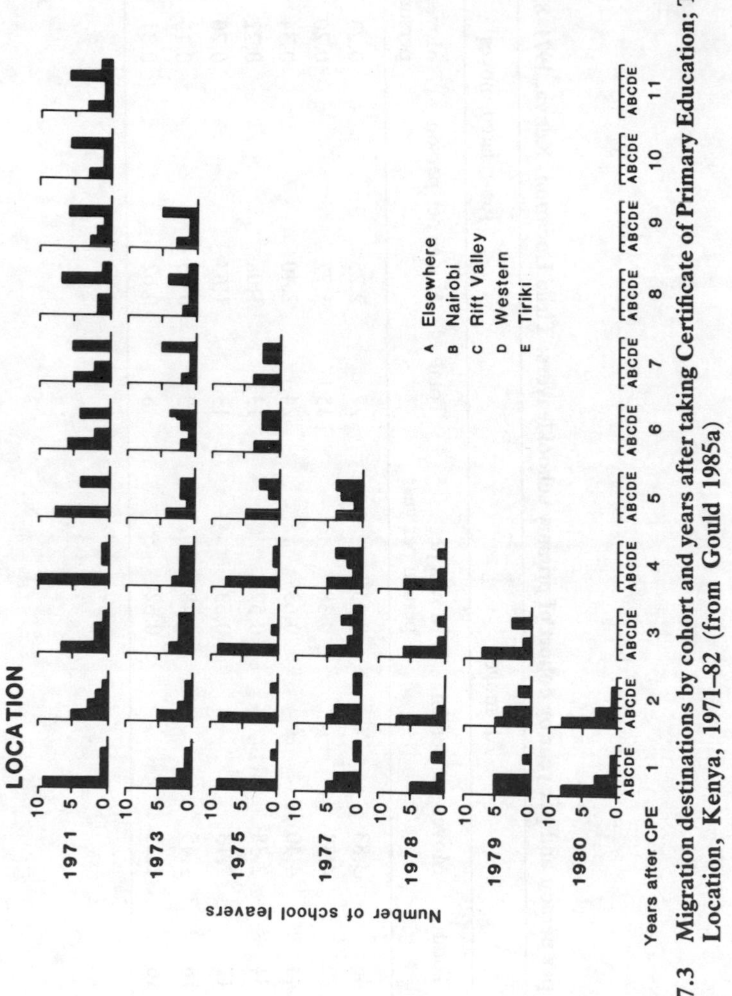

Fig. 7.3 Migration destinations by cohort and years after taking Certificate of Primary Education; Tiriki Location, Kenya, 1971–82 (from Gould 1985a)

compared with the distribution of their relatives in paid employment (Fig. 7.4(b)). Nairobi has the greatest attraction for the school pupils, with much lesser stated preference for seeking a job in more nearby locations, including in Tiriki itself. This pattern is rather different from that of the actual job locations of their relatives. The largest single place of residence of the relatives is Nairobi, but there is a much larger proportion in Western Province (Kakamega, Busia and Bungoma Districts) and in Rift Valley Province. For both groups proportions in Nyanza, Mombasa and the 'rest of Kenya' are similar and small. These data showed that the average educational level of the relatives was lower than that of the school pupils about to leave primary school, and that they were able to find jobs locally in rural agricultural jobs (e.g. as labourers on sugar estates) and rural off-farm activities (e.g. as drivers) within Western Province, often after a period of work or at least job search in Nairobi. The expectations of the school pupils are clearly biased towards the urban and distant destinations, and they may indeed migrate to them as a first migration. However, local destinations, even for these pupils, may be more likely in the longer term.

This relative stability and continuing out-migration of primary school leavers in the period covered by the retrospective tracer survey, 1971–82, has to be reviewed in the context of a continuing expansion of school enrolments, especially at secondary school level, and a sluggish expansion of job opportunities in towns over that period, circumstances that would be expected to impose further constraints on primary school leavers getting a job. But clearly for this area the expected response does not apply, for this is an area of heavy out-migration in the face of severe land shortage and limited off-farm opportunities for the more highly educated as well as for primary school leavers. Migration out of the area is an essential strategy for household survival for it generates remittance income. Education is a critical expenditure of each household for it produces a supply of migrants, regardless of the demand conditions in the urban labour market. In such an area Conroy's model does not seem to apply in its later stages.

However, this conclusion is regionally specific, for there is evidence to suggest that migration propensities of primary school leavers are lower in other parts of Kenya, and particularly in provinces with higher overall levels of income and educational attainment. Collier and Lal's (1986) conclusions about the importance of urban earnings for rural investment in Central Province, an area with a much more buoyant rural economy than Western Province, are that overall migration rates to Nairobi seem to be falling as the rural option becomes more attractive. This is further confirmed by the evidence of answers to an identical set of attitudinal questions first administered

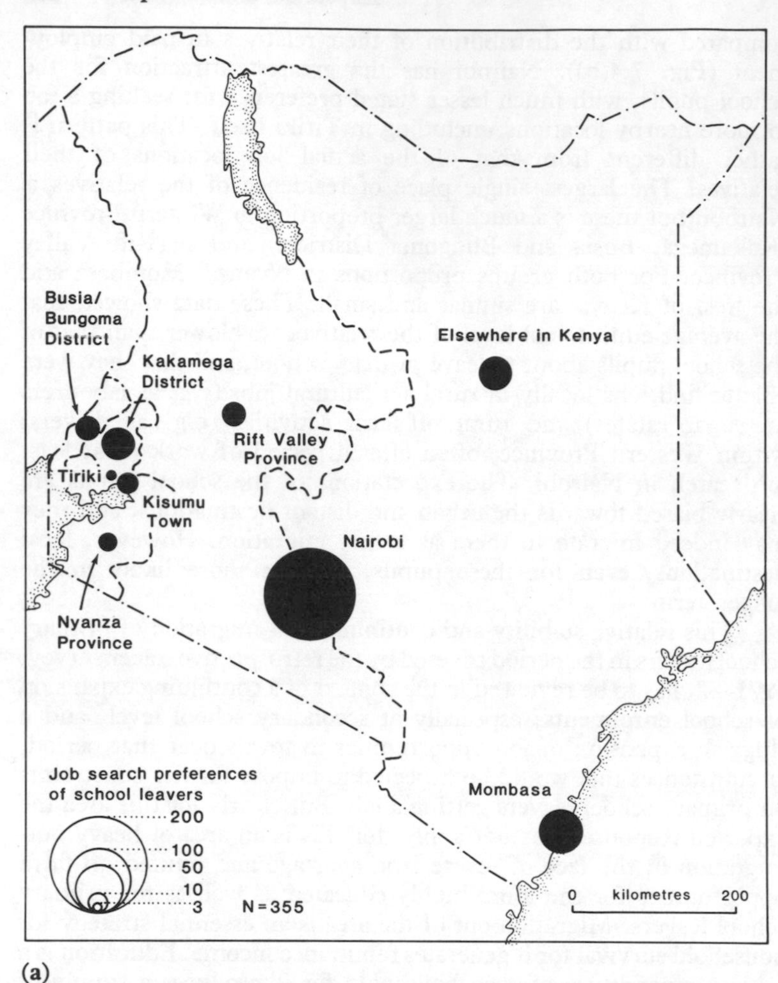

(a)

Fig. 7.4 Migrations from Tiriki, Kenya. (a) Migration preferences of final year primary school students.

to final year primary school students in Tiriki, Western Province, in 1981, and subsequently to a similar group in a village in Central Province in 1985 (Ekberg *et al*. 1985). Although interpretation is complicated by the different year in which the surveys were undertaken, it is clear that there are important differences in the attitudes of pupils in the provinces that are consistent with these other conclusions (Table 7.3). Pupils were asked to indicate whether they agreed or disagreed with three statements about migration choices. While there is relatively little difference between the two sets in preferences for seeking new land after completing

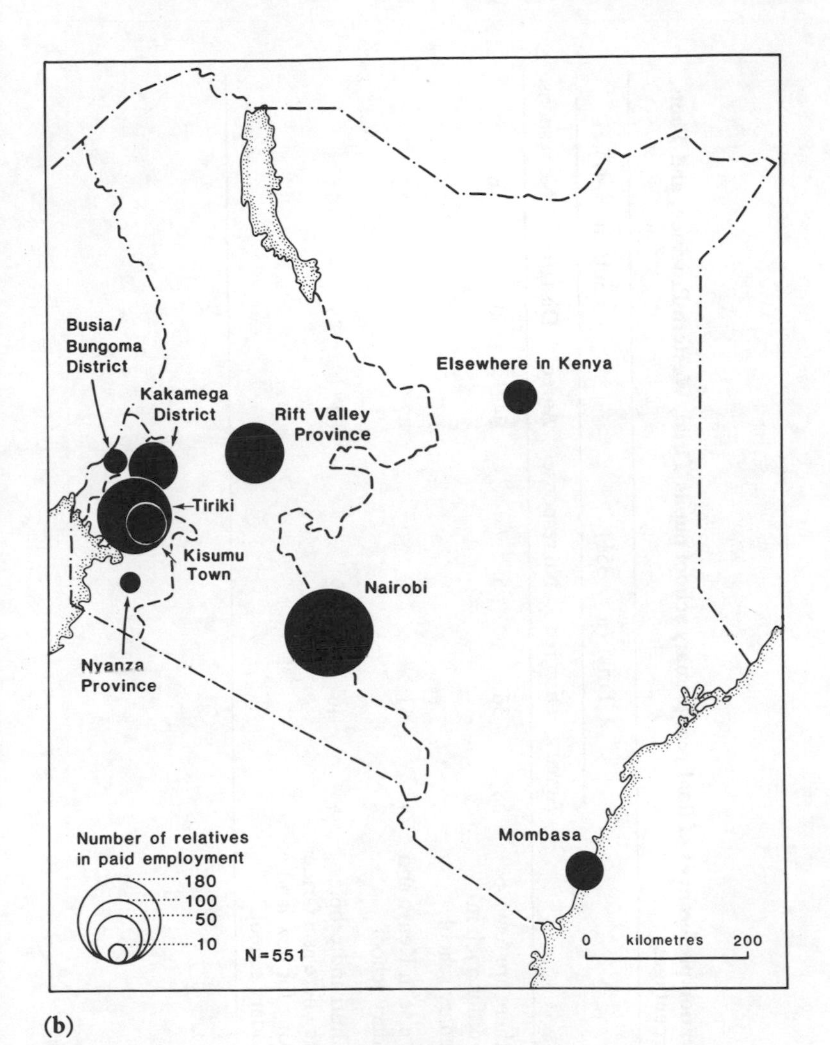

(b)

Fig. 7.4 (b) Current location of relatives in paid employment

primary school, pupils in Western Province expressed a much stronger preference than those in Central for migrating to town to look for a job; conversely pupils in Central expressed a much more positive attitude to local farming opportunities.

These regional differences in Kenya find an echo in Lawrence Brown's migration studies in several Latin American countries. He emphasizes the importance of place: that people with similar individual characteristics including educational attainment, may display different migration behaviour related to the place where they live, the basic geographical point (Brown 1990). Specifically in the case of

Table 7.3 Rural and urban migration preferences of final year primary school pupils Tiriki, Western Province, and Thuti, Central Province (percentages)

	Tiriki (n = 351)			Thuti (n = 47)		
	Agree	Disagree	No response	Agree	Disagree	No response
Although towns are crowded and there are few jobs and housing is expensive, I will look for a job in town if I fail to get a place in secondary school	58	36	5	34	60	6
I want to look for land somewhere else in Kenya and be a farmer if I cannot go to secondary school	43	51	6	42	45	13
Although land is scarce in Tiriki/Thuti and people are quite poor, I would prefer to stay here as a farmer to improve my land instead of looking for new land or a job in town if I do not go to secondary school	45	46	9	64	25	11

Source: Gould 1987.

Venezuela, he illustrates how attempts to spread educational opportunities in the 1950s resulted in increased migration from the previously underserved areas for any given level of education, for the increased educational opportunities in these regions were not accompanied by an equivalent spread of employment opportunities. Industrial expansion in the period and in the 1960s was closely linked to the Cuidad Guyana Complex and to urban areas near Caracas, the capital city: 'This difference in the scale of decentralization exacerbated spatial disparities between employment opportunities and the educated work force. . . . The net effect was to increase migration propensities already present in a highly polarized, urban focused economy' (Brown 1990:125). Furthermore, educational expansion in peripheral areas was taken up more by previously disadvantaged women than by men. Since the formal labour market for educated women was, even more than for men, concentrated in these urban areas, mostly in service employment, the migration propensities for female school leavers were raised even more than those of men.

These regional differences affect not only formal school leavers but other groups of young people, including youth polytechnic trainees in Kenya. As was discussed in Chapter 6, the youth polytechnic movement was established to provide rural youth with rural skills and so slow down, if not prevent, rural–urban migration of school leavers. However, it has largely failed to do this in Western Province where the well-established culture of migration has meant that the skills and attitudes acquired in the polytechnic have promoted migration. This has been shown also to be the case for polytechnic trainees from Central Province (Barker and Ferguson 1983). In comparison to the Kakamega case, polytechnic leavers in Central Province have lower rates of migration, despite nearness to Nairobi, for the trainees are more easily and more productively absorbed into the relatively prosperous rural economy, with greater cash demand for carpentry, and for house building and tailoring products.

In Kenya and Venezuela, as elsewhere, the nature of the education/migration relationship is clearly affected by local circumstances and is regionally differentiated within any one country. Overall there may be a weakening of the relationship over time, with a progressive flattening of the curve, such that for any given level of education a rising proportion of school leavers do not become migrants. However, where there are continuing rural–urban and regional differences in job opportunities and where, as a result of rising population pressure, there are very few local economic opportunities on and off the farm, the relationship may not become weaker over time.

The joint effects of differential provision in richer areas and differential migration of the educated to these areas produce sharply differentiated structures of the population by educational status. The extent of the differences can often be identified from census data. The Kenya Census of 1979 identified the population of each district by age group and educational status and the results for Kenya as a whole, for Nairobi and for Kakamega District are described in Fig. 7.5. The diagrams identify the relative recency of educational expansions, especially for females at all levels and for both males and females in secondary level. Over 50 per cent of the male population aged 20 in 1979 had had some secondary education, roughly double the proportion of those aged 30,

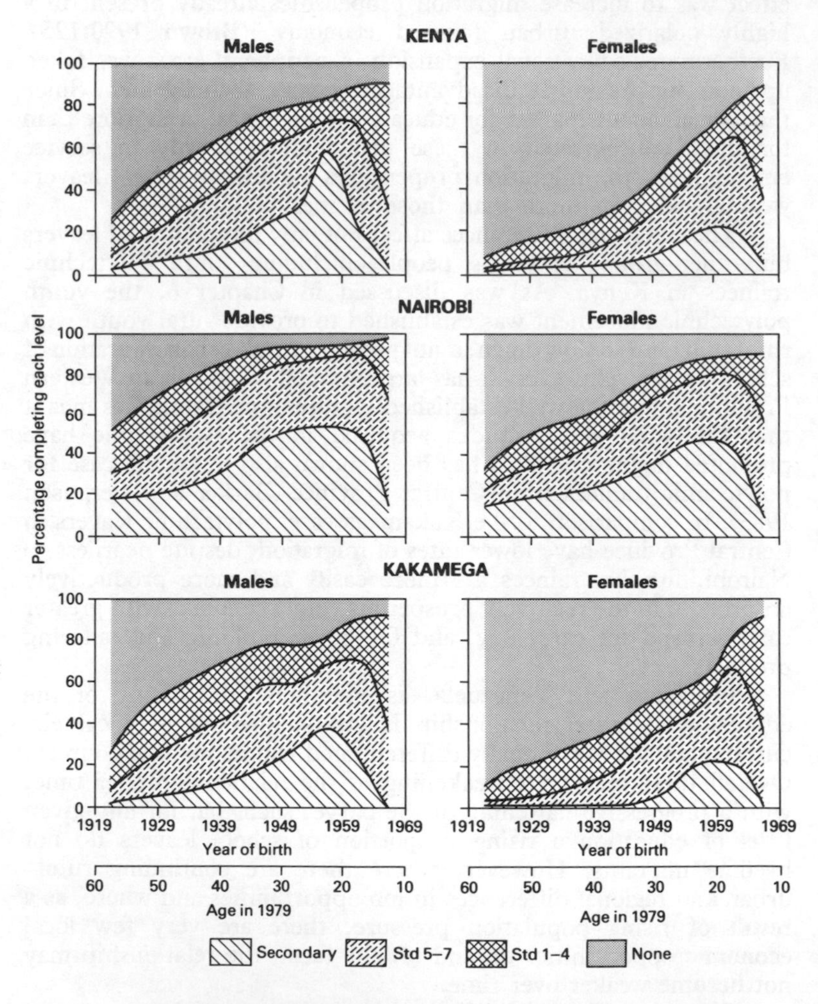

Fig. 7.5 Educational characteristics by age and sex: Kenya, Nairobi and Kakamega 1979 (from Gould 1986:191)

proportions that are well in excess of the proportions of females who had had secondary education. However, the key issue here is to note the differences between Nairobi and Kakamega, and particularly with respect to secondary education. While it is true that levels of secondary provision have always been higher in Nairobi than in Kakamega, the differences in the proportions of the male and female populations at all ages with secondary education is much more a result of migration of the educated to Nairobi – and staying there at least until retirement – and out of peripheral districts like Kakamega. Whereas over 40 per cent of the population of Nairobi in the age ranges 15–40 have been to secondary school, this figure exceeds 30 per cent only for the 20-year-olds and is only 10 per cent of those aged 40. In Kakamega the proportions are very low for those aged over 50, but in Nairobi almost 20 per cent of males aged over 50 have been to secondary school. A similar pattern exists for females, but with lower overall proportions.

The urban migration of the educated as revealed by the census is confirmed by the National Labour Survey 1976/77, which showed that while 11 per cent of the rural labour force had secondary education the figure was 37 per cent for the urban labour force (Bigsten 1984). The rapid growth in the public service as a proportion of all formal sector jobs in the years after independence in most Third World countries suggests that the Kenyan experience may be fairly typical, for the skills of the bureaucrat are those of literacy and numeracy, skills acquired through formal education. So called 'academic' education is in fact vocational in that it gives people skills that allow them to be eligible for highly prized jobs: relatively well paid, relatively secure, not particularly taxing for the majority, but giving access to urban life. There is a strong link between social and spatial mobility. The personal and family prestige associated with government employment is in part associated with urban residence. Even, as in recent years, in countries seriously affected by retrenchment in the bureaucracy as part of structural adjustment programmes (in Ghana for example, central government is likely to shed 50 000 civil servants, 1987–92) and even with salaries that are well below those which prevailed a few years ago, government jobs in towns are still sought by school leavers. Decentralization of government jobs has been high on the rhetorical agendas, much discussed in official policy documents, but has not generally progressed much in practice.

The migration of the educated to towns is largely due to job opportunities rather than rejection of rural life and values *per se*. It may, nevertheless, be affected by the organization, curricula and attitudes in the schools. National policies for the organization and management of the schools system may affect the propensity and timing of migrations of school leavers. Age at initial entry and age of leaving each of whatever cycles occur can mean that many school leavers may be too young to enter the urban labour market. They may stay on in school until they reach an appropriate age, regardless of real achievement during these extra years; alternatively, they may leave school but, if they do, will they simply postpone migration for some years or will some completely abandon it, given that they may find an acceptable rural occupation in the intervening period? This impinges upon the policy debate about appropriate ages of entry to and promotion within the system, and the desirable length (including the vexed issue of repeating) of compulsory or universal education. An earlier start, with a shorter basic cycle and no repeating, may result in delayed entry to the labour market and lower rates of migration in the longer term. Recent structural changes in many countries have recognized the need to organize systems such that the main points of exit are those at which urban migration is possible or even expected.

Furthermore, each school can also have an impact on the aspirations and expectations of pupils that is independent of the more general effects of the organization of the schools system. Chief among these in-school variables are teachers, both in so far as they directly affect achievement of pupils and indirectly as they affect their wider attitudes to local society and the state. In both respects teachers can affect migration behaviour of school leavers. Teachers' performance in the classroom clearly affects individual achievement. In some countries a posting to teach in a rural school is considered by teachers to be a punishment posting, or at best a necessary evil to be borne by a young teacher before he or she be appointed to an urban post. The career mobility of school teachers and their desire for an urban post may in itself affect and be reflected in the migration aspirations of their pupils. All too often the teachers come from urban areas, and do not become integrated into the rural community, and as representatives of the 'educated' are seen to be different, essentially urban people.

The long trend in most, but not all Third World countries, is towards considerable urbanization, and the migration of the educated certainly highlights the trend. This has raised the fear in some quarters of a long-term genetic drift of the intelligent out of the rural areas into towns with adverse long-term implications on those left behind and for future generations. Such a view can be largely rejected for two main reasons. In the first place there is no necessary link between education and innate intelligence, and particularly so where schooling is not compulsory, and attendance, especially at higher levels, is much affected by social background. Secondly, since there is still a lot of rural–urban interaction and the educated may find marriage partners (or have their partners found for them in arranged marriages) in rural areas, then the drift will be slowed down if it operates at all.

More generally the term 'internal brain brain' has been applied to the general pattern of movement described in this section, but with a presumption that the movements exacerbate rural–urban differences, and further contribute to rural underdevelopment. Its use in an analysis of 'internal brain drain and development' in Senegal invokes explicit consideration of higher education. Since higher education students are disproportionately from urban backgrounds in the first instance, future urban-biased residential preferences are hardly unexpected (Johnson 1985). However, students from small town and rural backgrounds and from low socio-economic status households display equally strong anti-rural preferences. In this case the surveyed students, many of whom were already educational migrants, were studying in the capital city. The urban location of institutions of higher education is a further contributor to the bias in students' perceptions of their future preferences and prospects.

The issue of internal brain drain assumes greater significance in the larger and more complex case of Indonesia. As a contribution to the formation of a province-specific population policy for the peripheral province of North Sulawezi, Gavin Jones has used the term to express concern for an internal loss of locally trained skilled professionals to Jakarta (Jones 1989). In this case the local opportunities in higher education create a supply of professionals that is larger that the local job market can absorb, and the graduates seek and are more likely to find appropriate jobs in the national capital. North Sulawezi experiences key manpower shortages as a result of the migration.

To conceptualize the migration of the educated as an 'internal brain drain' may be misleading, and especially since there is strong evidence that urban areas and the people working in them can have a strong and direct effect on rural development through continuous urban–rural interactions, movements of goods, capital, information and ideas as well as people (Gould 1988a; Potter and Unwin 1989).

International migration: the brain drain

The brain drain is a term more commonly associated with international migration. It was first coined in the United Kingdom immediately after the Second World War to identify the migration of skilled British scientists to the United States, but by the 1960s had been generalized to a global phenomenon and particularly to the migration of skilled professionals – doctors and engineers in the main – from Third World countries to First World countries, predominantly to North America, but also to Western Europe (Glaser 1978). It is a phenomenon associated with education, and in particular with the very highly educated. The most highly educated, as has been discussed above, display the highest propensities to become migrants and are most likely to move over long distances internally within any country. However, their education and skills allow them to be seen as being part of an international, perhaps even global, labour market. Skilled professionals are willing to cross national political boundaries; foreign countries arc willing to accept these workers into their own labour force. In one sense therefore the international migration of the highly educated is not different in essence from that of internal migration, with the same sort of differentials and impacts at source and destination and raising similar sorts of issues about the amount and type of education as a cause of movement, and of policy measures to reduce the deleterious effects, particularly at source.

In another sense, however, there are fundamental differences between the internal and international migrations that require distinctive and separate treatment. With internal migration the redistribution of the benefits of the migration, benefits which accrue to the individual and his or her family as well as to the society and economy as a whole, is relatively easy or at least facilitated by the state through, for example, extending social facilities into peripheral regions from which some of the migrants come. Even if the benefits accrue disproportionately in the destination areas, and principally in the capital city, there are mechanisms which can be established to redistribute these benefits, though these mechanisms are seldom as strongly implemented as they can be or are claimed to be. In some cases they may hardly be invoked at all, but that is a matter of internal political choice. At the international scale, however, redistribution of the benefits that disproportionately accrue in the destination countries is so much more difficult. Redistribution by government to government through aid and other trading arrangements is a small proportion of income generated, and has been declining in recent years. Much attention has been directed to the redistributive effects of individual remittances on the source economies and the extent of their impact is a matter of very considerable controversy. They can be substantial in volume, but tend to be used for consumption expenditure at the origin (larger and better housing, consumer goods, especially imported electrical goods, cars and lorries, etc.) such that their impact on local investment and development at source is limited (Keely and Tran 1989; Looney 1989). This is a general issue affecting migration of unskilled labourers – the 'brawn drain' – as well as skilled workers, but can be seen as consistent with a global process of underdevelopment in which the development of rich countries is supported by and at the expense of resources – in this case labour – from the Third World.

Contemporary patterns of international migration certainly seem to exacerbate and perpetuate global inequalities, and to warrant the ascription 'drain'. Migrations are numerically dominated by the unskilled – from Mexico into the United States, from Mediterranean countries into Western Europe (the *gastarbieter* phenomenon), from Arab and South Asian countries into the oil economies of the Persian Gulf – a phenomenon of workers to the work (Salt 1989). Though in the new international division of labour there is also some transfer of work to the workers, notably through the mechanisms of multinational companies and focused particularly on the countries of East Asia, from Korea south to Indonesia, this seems to exacerbate global inequalities (Thrift 1989). The highly skilled constitute an easily identifiable and important group within the broad spectrum of migrants, and their qualitative if not numerical importance has made them an issue of particular concern to policy makers and commentators on development, labour force formation, migration policy and education (Gould 1988c).

It was not until the late 1960s that migration of skilled professionals from the Third World to the First World became a widely discussed issue. Its rapidly growing importance from that time is associated with changing immigration criteria in many First World countries, moving from racial or national origin criteria to economic criteria. Immigration policies became strongly linked to economic policies and to internal labour market conditions. Nowhere was this more obvious than in the United States, where the 1965 Immigration and Nationality Act largely abolished national quotas for entry to the United States, but introduced economic criteria that specified admission for those with specific skills, and in occupations that were in shortage in the United States. The category of skilled professionals is separately identifiable in US immigration statistics and permits separate identification of social scientists, engineers and physicians. Since 1972 over three-quarters of all professional immigrants have come from 18 Third World countries for which data are separately available (Table 7.4). The table illustrates how small is the professional category (*c*.5 per cent of the total), but underestimates the full extent of the brain drain in two important respects. Firstly, many skilled workers are in other categories: e.g. nurses, teachers, technicians, and those are significant groups (Rockett and Putnam 1989). Secondly, the 18 countries included as Third World are the largest contributors in gross terms, and include eight Latin American countries, one European country (Greece), nine Asian countries, but no African countries. Given that there are over 100 Third World countries, the majority of them make only a small contribution to the US total. However, they may comprise a very large and certainly very significant proportion of skilled workers in these smaller countries.

While legislation and immigration policies altered the structure of demand that resulted in increased immigration of skilled workers to the United States, Australia and the European Community, perhaps more directly relevant to the discussion of the relationships between education and migration have been factors in the Third World countries that have increased the supply of highly educated workers in skilled occupations. Three factors are particularly relevant here: over-expansion of higher education provision, internationalization of qualifications and student migration.

As was noted in Chapters 2 and 3, there have been massive expansions in education in Third World countries throughout this period, with greatest expansion in proportional terms at the higher level. Those expansions were planned in the 1960s or earlier to ensure an adequate supply of graduates some 15 years later from an

Table 7.4 Trends in United States immigration 1972–87 (from MacPhee and Hassan 1990:1112)

Year*	Total immigration	Total Third World immigration	Per cent	Immigration of professionals Total	Immigration of professionals Third World	Immigration of professionals Per cent
1972	384 685	281 868	73%	18 466	15 684	85%
1973	400 063	296 352	74%	13 706	11 321	83%
1974	394 861	304 294	77%	10 902	8 675	80%
1975	386 194	303 390	78%	12 784	10 270	80%
1976	398 613	316 768	69%	14 203	10 906	77%
1977	462 315	377 631	82%	15 554	11 827	76%
1978	601 442	509 196	85%	15 360	11 497	75%
1979	460 348	383 732	83%	NA	NA	
1980	530 639	442 700	83%	NA	NA	
1981	596 600	516 767	87%	NA	NA	
1982	594 131	512 162	86%	12 200	9 316	76%
1983	559 763	487 627	87%	10 500	7 798	74%
1984	543 903	467 133	86%	9 600	6 910	72%
1985	570 009	493 540	87%	10 900	8 075	74%
1986	601 708	526 193	87%	11 200	8 185	73%
1987	601 516	523 318	87%	11 300	9 169	81%

Source: United States Department of Justice (1987). National Science Foundation (1986).

* 1972–78: social scientists, natural scientists, engineers and physicians.
1982–87: social scientists, natural scientists and engineers.

initial cohort entering primary school. The long gestation period while cohorts of the required size were progressing through the system was built on an assumption of rising levels of economic activity with increasing demands for skilled manpower. However, the optimistic projections for manpower needs made in the 1960s were seldom realized by the 1980s, so there seemed to be a surplus of skilled manpower. It was most acute in countries with the most dismal economic performance.

In Africa it affected Ghana and Uganda more than other countries. Both these countries in the colonial period had relatively well-developed schools systems and at independence in 1957 and 1963, respectively, they were both richer than neighbouring countries. They seemed set fair for continuing economic growth and developed expansion plans in education that sought to meet the substantial manpower needs of their richer and expanded economies. However, in both cases there has been severe economic collapse for largely internal reasons: due to political instability in Ghana, and the catastrophically destructive effects of the Amin regime in Uganda, 1971–78. Internal demand for local graduates collapsed, but the local universities kept on producing them. The net effect was massive out-migration of educated workers to neighbouring countries. Nigeria was the major destination for Ghanaians, but with major political implications for Nigeria that prompted the abrupt expulsion of many of the migrants in 1983. Ugandan graduates, and particularly medical doctors from the prestigious Makerere Medical School, are very widespread throughout Africa (Gould 1985c). Clearly not all African countries have suffered economic decline as severely as Ghana or Uganda, but the differential rates of economic growth coupled with differential rates of educational expansion (relatively low in Nigeria and in the countries of Southern Africa coming later to independence) produced a more localized mismatch between supply and demand that created conditions for increased migration of skilled manpower within Africa (Gould 1988c).

This more localized intra-continental brain drain has a parallel within Latin America to a lesser extent, but hardly at all within South and East Asia. There is substantial migration of skilled professionals between countries of the Arab world which is similar to more general patterns of unskilled migration from labour surplus to labour shortage countries of the region, with Egypt as a major source of professional and skilled manpower. It has had a larger education system for a longer period of time than most of the Gulf Arab states, and has therefore larger stock of educated workers. This also applies to Palestine and Jordan, and in this case the emigration of skilled workers has been accelerated by the political problems over the existence of the Palestinian state. In these circumstances there is a tendency to see the migration as less of a 'drain' and more as an 'exchange' of manpower, for the moves are often temporary. Benefits that accrue within the country of destination at least accrue within the Third World, and there is a regional expression of benefit and relatively easy mechanisms for remittance of individual savings if required. A local surplus of skilled workers need not necessarily lead to migration out of the Third World, but can contribute to Third World development elsewhere. The United Nations and other agencies, such as the International Committee for Migration (ICM) have sought to encourage brain 'exchanges' within the Third World, but these have been very limited in their scope, and have had even less impact than parallel schemes to encourage return migration of skilled workers from the First to the Third World (Keely 1986).

A second supply factor concerns the international standardization and acceptability of qualifications. International comparison of some qualifications can be extremely difficult, and quantity and quality of achievement can vary very considerably for any given qualification from one part of the world to another. In some professions and for some countries there may be standardized qualifications, for example between many British Commonwealth countries. This is even more common practice within French Community countries. In other contexts, and particularly for the United States, foreign migrants are required to take professional competence tests as well as language tests as part of the immigration process, in some cases in the country of origin. The path to migration of skilled workers is smoothed by increasing use of standard tests and international recognition of formal qualifications.

A third and probably much more important factor in the brain drain and one which relates most directly to migration from Third to First World countries is the fact that many Third World students study abroad, mostly in North America and Europe, and are financed by scholarships from First World countries or institutions. They attend specialist courses, mostly at a postgraduate level, and gain qualifications that are not only internationally recognized, but allow easy, often immediate access to the labour market in the country in which they have studied. Employers are more sure of the range of competences they have, they can interview applicants whose skills are more likely to be related to practices and current levels of technology in the countries concerned. They will be more directly able to satisfy immigration requirements for shortages in specific jobs. Costs and risks to employers of employing locally trained and immediately available graduates from Third World countries are therefore reduced.

Even more critical is that the student in question, having spent some time in a rich country, enjoying a standard of accommodation and current consumption and future prospects that he or she will be unlikely to find at home, may prefer to find employment in the First World, at least for a few years, instead of returning home. The attraction of remaining is also a professional one. The graduate will know that a working environment as conducive or challenging for a keen young professional will prove difficult to find. And of course salary levels are likely to be much lower in the Third World. For social, professional and economic reasons there may be a strong incentive not to return. There may also be social difficulties, notably over marriage and conflicts with families at home, but these can be overcome by a spouse from home joining the graduate in his new place of work (Helweg 1987). Most students, however, do return, often unwillingly if they cannot find a job with immigration permission, but many do remain to contribute to a permanent brain drain.

In general the issues raised above confirm the view that the brain drain has to be seen as one symptom of broader processes of peripheralization of the Third World as a result of unequal exchange. There is a large and growing literature on the broader phenomenon of contemporary international migration, with consideration of problems such as patterns and trends (Salt 1989), experiences on migration and development (Appleyard 1989), and evaluations of the role of remittances (Keely and Tran 1989), and into which the specific issues associated with the brain drain fit. The literature, notably the journals *International Migration* and *International Migration Review*, deals with issues in the countries of origin, e.g. India (Ommen 1989), Sri Lanka (Eelens and Speckmann 1990), or specific groups at destinations, e.g. Korean physicians (Shin and Chang 1988) or Filipinos (Liu *et al.* 1991) in the United States, foreign teachers in Canada (Wotherspoon 1989), and South Asian women in Kuwait (Shah *et al.* 1991), and considers the very specific problems of the highly educated migrants in their own right or as part of a broader consideration of international migration.

It is necessary in this context now to turn to the question of how education affects these skilled international migrants. As with internal migration it seems analytically impossible to separate the 'education' factor in the migration from other related factors. A high level of education is not the sole 'cause' of the migration. Education along with other factors at source and destination creates differentiation of some potential and actual migrants from others. It affects the career and assimilation patterns of those who move. What is important is to consider how the quantity and type of education provided at the higher levels can affect the volume, direction, periodicity and economic significance of the movements in this broader context of structural inequalities between source and region that seem to dictate the economic bases of the broad pattern of flows and their effects.

The quantity of education provided has previously been raised. Although it would appear that the number of migrants is magnified by over-production of highly educated manpower at very high cost, to an extent that the Third World country cannot internally adequately absorb these workers, this is misleading. There is self-evidently a shortage of doctors in India, of engineers in Peru, of scientists in Senegal and of graduates throughout the Third World. The needs are great and productive possibilities for application of the skills of a greatly expanded highly skilled labour force are everywhere. The migrations occur because not only is there inadequate finance to employ as many graduates as the needs of the economy and the population would suggest, but because there is also a relative shortage of doctors in the United Kingdom, of engineers in the United States, of scientists in Sweden. The education systems of First World countries have not expanded sufficiently to produce the required manpower locally for the now much richer and much more technologically advanced economies than a generation ago. In the short term certainly, but it seems too in the longer term, it is preferable to import skilled workers rather than to educate and train locally the extra skilled workers that are needed. Costs of employing foreign graduates who have been trained in the Third World country at that country's expense (wholly or partially) seem to be less than the costs of investing in further education and training by First World governments and companies. Education and training policies in Europe and North America need to be considered as well as these policies in Third World countries, and should include issues associated with policies that further facilitate brain drain migrations through promotion of scholarship programmes in First World rather than Third World universities (King 1991).

The essence of formal education is that it is universal, and based on the Western experience and models. It is therefore internationally transferable in principle, and the educated acquire internationally transferable skills, like literacy and numeracy and basic scientific knowledge, or are sufficiently skilled in one language, for example, that they can relatively easily learn another if required. The school curriculum, especially at lower levels, may have a national 'relevance', such as an emphasis on agriculture, but that local relevance is subsumed within the broader acquisition of more universal knowledge and skills. The higher the level of education, the more universal it is likely to be and the more immediately transferable from one country to another. The general trends in Third World countries at secondary level and at higher levels have moved away from specialist technical and vocational education to a more diversified general curriculum which is not only much cheaper to provide, but in a very rapidly changing technology will permit a very necessary flexibility to suit new labour market opportunities for school leavers (Psacharopoulos 1991). And these can be sought abroad.

The transferability is even more apparent in university and other institutions of higher education, for both technical and non-technical subjects. Clearly some technical specialisms are more transferable and less culture bound than others. Basic sciences such as engineering or allopathic medicine require acquisition and use of universal concepts; social workers or lawyers or economists or geographers have much more locally specific requirements of their training, yet even here there are general principles that can be applied elsewhere. Poor countries with poorly equipped universities may be unable to provide a state-of-the-art training in computing or other technologies, but can normally provide enough of a basis for further on-the-job training for the most able students, perhaps in a First World context where the appropriate new facilities and instruments are available.

The problem of transferability of skills and technological sophistication of training is particularly acute for medical doctors and other health care professionals. The general trend in health care provision in the Third World has been towards low technology inputs associated with providing basic health care for the majority of the population (Phillips 1990), rather than in providing high technology curative services in specialist hospitals for the few. Training needs for physicians might be seen to be very different from those in First World countries. Yet medical schools in the Third World continue to provide students with high-technology skills as far as possible, as well as restructuring curricula to make the training appropriate to a basic health care approach. This is achieved by maintaining a technically sophisticated component and giving medical students skills and knowledge they may not normally expect to be able to apply in their own countries. Such training may stimulate a professional need for migration to a more satisfying hospital environment where the technology is available. There is considerable evidence to suggest that, in Africa at least, governments accept that a high proportion of local medical students will migrate after graduation, motivated by professional curiosity and satisfaction as well as by the very large salary differentials (Ojo 1990). Governments find themselves caught in the horns of a dilemma. If they raise salary levels for doctors and other health care professionals to internationally competitive levels they may reduce the emigration of skilled personnel, but at great and probably politically unacceptable financial costs. If they were to be provided with a level of medical technology that professionals would wish to have available in order to do their job to the best international standards implied by their training, the costs would be much too high. A substantial migration of some skilled staff may be the price that has to be paid if a basic health care policy is to be pursued in the country.

In other contexts, however, education and training may be provided explicitly to stimulate emigration. In several Middle East and South Asian countries, for example, remittances from migration have become critical to national economies. Governments have sought to promote immigration by providing training in skills needed in the countries of immigration, and specifically in construction trade skills such as those for draughtsmen and electricians. South Korea has created special technical high schools explicitly to train migrants (Hong 1983). Pakistan, amongst other countries, has sought to develop schemes for training artisans on the assumption that most of them would seek work in the oil-related developments in the Persian Gulf, but that the skills they acquire and subsequent experience gained would eventually be valuable in Pakistan when they returned (Tsakok 1982). Jordan, even more dependent on workers' remittances, has pursued similar policies to include training at university level (Appleyard 1989:491).

More commonly in countries traditionally dependent on migration and remittances of skilled as well as unskilled workers, educational provision is implicitly rather than explicitly used by government and by migrants and their families as a means to encourage circular mobility, with successful emigration and subsequent and even frequent return. In Caribbean islands, for example, the numbers of skilled migrants have grown rapidly in most countries since the 1970s, but their migration should not be seen primarily in terms of labour force losses. They generate remittances at higher levels than less skilled migrants that help sustain the household economy, and ensure sufficient support for an educational base for the next generation of skilled migrants (Thomas-Hope 1988).

Where there have been transfers of industry to Third World countries by multinational companies (MNCs) rather than of labour from Third World countries, the companies may need to provide a training programme for the local production workers they employ. However, they may also facilitate transfer of their own skilled managerial and technical staff, increasingly themselves from other Third World countries, as part of the internal labour market of the MNC (Salt 1988). This is a currently small, but growing phenomenon and is facilitated by company-specific training policies and management methods that are dependent on having a cadre of international staff that has sufficient basic education on to which the common company training can be grafted. Increasingly, the labour market within and between companies for highly skilled workers is an international one and the Third World is not isolated from the global trend.

Chapter 8

Models of education and models of development

This book has identified a wide range of experience in the development/education relationship. Just as 'development' has a range of meanings, as an objective as well as a means for economic and social improvement in human societies, so too has 'education'. Education can be measured by the amount of learning or by mere attendance in a school; it can encompass a wide range of features in its content and organization, in terms of the length and breadth of the student's experience in school and the impact of that experience after he or she leaves school. As a result it has a great variety of interactions with the development process and at all scales. Education affects people in the local community and in society at large in different ways. Conversely, the style and purpose of development has an effect on the education system. There is no simplistic, universal education/development relationship. Education has been taken to be an integral feature of development processes, with measures to expand the quantity and improve the quality of schooling as typical components of national development strategies. More development has seemed to require more formal education; just how much more or with what form and content has varied enormously from one developmental context to another. In different countries at any one time, and at different periods in the same country, the contribution of schooling to development has varied.

Education has been an issue of great concern to governments and to the peoples of the Third World. Their heavy investment in education has brought benefits to many individuals and to their communities. These benefits are experienced in higher standards of living and more secure economic futures, and the benefits of the education experienced by those who have been to school are everywhere apparent. For others, however, the attractions and benefits of education have proved to be illusory. Education has not brought them higher standards of living, nor has it guaranteed them a more secure economic future. The investments and sacrifices made by many families in the anticipation of benefit, directly to those sent to schools and indirectly to their families, have all too often had disappointing outcomes, leaving the educated person frustrated, disappointed and often less willing than he or she might have been without the school experience even to participate in the family or village economy. Demand has been great, but the total amount of economic benefit and its social distribution have not been able to match the demand, either in terms of specific job creation or in terms of general improvements in income levels. Some have been to school; others have not. Some have benefited from the experience; others have not.

Education seemed to be a key, perhaps even the master key, to Pandora's box of development. However, once the box had been opened the contents that were found in it were not sufficient to go round. Cynically it might be argued that education has held out to eager populations an unrealizably optimistic prospect of social and economic progress. This is a prospect that people continue to seek even where the conditions suggest falling opportunities for the educated to achieve the promised and much sought-after benefits. Education has perhaps become a drug or placebo that the mass of the people needs to believe in in order to provide itself with the possibility of escaping from the burden of poverty. Without it, all is lost, but with it only some can benefit. Most people seek solace in that prospect, for they may be among the lucky few. This perspective suggests, echoing Illich (see Chapter 1), that it is education that has become an opiate for the people of the Third World, dulling the sense of degradation, disappointment and underdevelopment, and encouraging a sense of false optimism in its potential as a vehicle for radical change and a false promise of future prosperity. For others, however, and echoing the arguments of the human capital school of economists, the contents of the box have proved to be a stimulant rather than an opiate, boosting performance and allowing international competitiveness, raising the sights of the population as a whole. These sights may be raised to perhaps unobtainable goals for which further doses of the stimulant have seemed in the past, and will continue to seem, to be necessary if the goals are to be achieved.

This text has been able to illustrate some of the ways in which education has proved to be a mixed blessing for those who have been to school, for their communities and for their countries. Formal education has improved the level of knowledge and skills of the population, the quality of the human resource base. It has given some people access to well-paid, modern sector jobs, and, less directly, has allowed people to exercise more choice over their lives. Nowhere is this more apparent than in the ability of educated women to exercise more choice over their own fertility. It has not, however, been able to ensure that these new levels of knowledge and skills could always be put to sufficiently productive use, either in rural or in urban areas or through migration, or that the individuals were suddenly liberated from a range of social and cultural constraints. Education has brought universalist ideas and knowledge of science and technology to many people, but has been unable to ensure that new techniques and machines that put that knowledge to use were available. Development has brought Third World countries into the global arena of education, through technical support agencies like UNESCO and through financing agencies like the World Bank, but has been unable to ensure that foreign advice and finance have not led to the uncritical adoption of inappropriate models of education and development generally from, or of relationships with developed countries. These could maintain continuing relationships of unequal exchange that give the Third World its cohesiveness as a meaningful generalization. Examples of such inappropriate relationships are evident in the external brain drain of skilled workers from Third World to developed countries, but also within countries as a result of increasing educated unemployment overall and increasing regional and rural–urban divergence in the benefits of the educational experience. Divergence in educational achievement and performance seems to be broadly symptomatic of the more general experience of the last decade of growing spatial inequalities in levels of development at most scales.

Though education *can* be a liberating force, opening new perspectives and opportunities in social and economic spheres to many poor and also many not so poor people in poor countries, in practice it seems to have been more commonly a force for stability and the *status quo*. The greatest demand has been for the 'white-collar' skills of the bureaucrat and civil servant, groups for whom education has brought relatively high rewards throughout the Third World. The bureaucratic state, often a former colony, has tended to reproduce its own political, administrative and social structures through education. It has not been able successfully to use the schools system to generate economic dynamism through technical skills appropriate to an industrializing country or to a country anxious to adopt new technologies as part of structural economic change. Nor has it been able to generate through the schools social change appropriate to a poor country with large and often growing social inequalitites. Where there have been innovations in 'relevant' education they have had, at best, limited impact, and 'experience indicates that relevance programmes for school-going children are *not* an appropriate vehicle for the rectification of social and economic ills' (Sinclair with Lillis 1980:163).

Those relatively few countries that are experiencing major economic transformations have not seen that transformation as being led by educational reforms. In some of the NICs the education system has been altered to provide more support for technological change, and nowhere more so than in South Korea. The rapid transformation of the Korean economy since the 1950s, driven by a central economic planning strategy, required a very rapid increase in scientific, technically skilled workers (Hong 1983). These would be produced in part within companies, but the Government, particularly from the 1970s, invested heavily in four very specific types of technical high schools:

1. machinists' high schools training machinery production workers, which between 1973 and 1981 produced 21 900 of the nation's precision workers;
2. technical training for work in construction projects abroad, with an output of 2000 students per year in the early 1980s;
3. specialized training in electronics, chemical, construction and railroad industries, industries that were expanding rapidly but had experienced critical manpower shortages, with some of the costs of these schools being shared between government and the companies that would benefit;
4. general technical high schools.

The economic transformation of the country has required restructuring of education towards technical training, but the expansion of technical education has tended to follow rather than lead the economic restructuring. In Singapore, too, the role of the human resource base has been identified as a critical feature of Singapore's economic success, with, as in Korea, considerable training collaboration between government and the private sector. Chew (1986:133) reports that 'management in the electronics industry generally places a lot of emphasis on up-grading of human capital'. The role of formal technical education as a factor in the economic dynamism of East Asia is contributory rather than dominant, and certainly much less than the role of foreign capital and internal political stability have been.

However, in the mass of countries in South Asia, Africa, Latin America and the Middle East there has, on the whole, been 'more of the same'. There have everywhere been changes in curriculum, structure and finance, but with relatively few attempts to alter attitudes to education fundamentally, either from within or from outside the education system. The education system has seemed to ensure social reproduction rather than be a vehicle for economic dynamism or social change. Even where the national state has in practice collapsed, the education may continue to function. In Sierra Leone, for example, now the poorest country in the world with the lowest score on the UNDP Human Development Index (UNDP 1991), the schools system remains in place, but its quality has seriously collapsed at all levels. Teachers are very irregularly paid what are in any case extremely meagre and much devalued salaries, so that they must spend most of their time outside school earning additional income in the informal sector. Buildings decay, there is no expenditure on learning materials, and generally the schools system mirrors a nation in extreme crisis (Banya 1991).

Political leaders in the Third World have seen to it that education has usually been consistent with national ideologies for development. Education has been centrally integrated into 'development planning'. It is prominent on the international development agenda, an area of government and private sector activity that attracts much aid and technical assistance as well as academic discussion. However, national and international development agendas have tended to be dominated by the unequal global power and trading relationships that are at the heart of the persistence of the Third World as an idea and as an objective reality. Initiatives in education, however well intentioned and however well supported politically and financially, from within and from outside each country, can only operate within the confines of the broader relationships. This has not meant that all countries have been affected in the same way. Far from it, as we have seen. The diversity of the national experience is indicative of the range of efforts and approaches to educational change. Countries have not been passive participants in a global educational agenda. They have had their own priorities and objectives – formal schooling and/or training; primary/secondary or tertiary level; rural and/or urban skills; state and/or private schools. However, these have only been able to operate within the internal parameters set by national political goals and national development strategies and the externally imposed constraints of structural adjustment and growing indebtedness.

Most of the discussion in this book has used the terms 'education' and 'schooling' almost interchangeably, and attention has been directed to the formal education system in particular, for it is formal schooling that presents governments of Third World countries with major policy and resource dilemmas, and it is formal schooling that consumes most of the public and private resources allocated to human resource development. It is school education that is the focus of the strong demand by parents and children. Nevertheless, it needs to be recognized that education is broader than real schooling, and that many of the relationships discussed in Chapters 5–7 of the text, e.g. between fertility and mortality and education, between employment and education, between migration and education, merely use schooling and achievement in school as an easily measurable proxy for a wider range of attitudes and levels of personal maturity and competence. Schools can certainly add significant value to these levels of maturity and competence, but the general weight of the evidence on the education/development relationship seems to suggest that the value added by formal schooling may not be as great as economists' rate of return calculations might suggest, and that generally the direct effects of schooling may have been overstated in the economic models of development invoked to underpin policy strategies of governments.

In political terms the necessity for expanding school provision to sustain democratic legitimacy has kept formal schooling to the fore. For governments of the Right, e.g. in Pinochet's Chile (Aedo-Richmond *et al.* 1985) or in South Korea (Hong 1983), investment in schooling has sought primarily to support modernization in the economic sphere and to maintain social stability. For governments of the 'liberal' Left and Centre, e.g. Zimbabwe (Dorsey 1989), it has sought to both promote economic change and social equity, but with an emphasis on the latter. For socialist states education and schooling have been seen to be at the leading edge of social and economic change, e.g. in Nicaragua (Doherty 1988), Ethiopia (Negash 1990). In each of these groups of countries formal schooling has been seen as the principal means of government intervention in education, but nowhere has it been given the lead role relative to other developmental factors such as level of investment, quality of transport, availability of natural resources. It is one of a range of factors, of potentially greater importance to the overall strategy in more populous and poorer states than in large countries with variable but relatively well-developed natural resources but relatively small populations. But nowhere, as indicated by levels of government expenditure, has education been reduced to an insignificant place relative to those other factors.

All these countries, while internally following strategies based on widely differing assumptions, have been caught up in a world system that imposed constraints on educational priorities, but has implicitly assumed a universalist and diffusionist model of development. In education it has implied an assumption, fully borne out in the operation of multilateral and bilateral aid agencies, of Third World countries being positively affected in their development by the operations of the development/education relationship that seemed to apply in Europe, North America and Japan. Education systems and institutional structures, and emphases within them that had been developed in these richer countries, would be appropriate for Third World countries. Just as in the rich countries the education systems were different in detail, and displayed different emphases and priorities, so too they could in the Third World. However, the central core was not to be denied: that of emphasis on a formal schools system, centrally planned and managed, with assumptions and objectives consistent with the broader national development strategy.

In this respect, models in education have been similar to

models in development generally. Hettne (1990) has argued that development strategies in Third World countries have been characterized by three basic features: they have tended to be exogenously based; they have tended to be rooted in a positivist ideology derived from the rationality of classical economic theory; and they have been essentially evolutionary, following the presumptions of a Western development model based on stages of growth. The assumption of an approach to global development using universally applicable theories and practice has been a characteristic feature of the second half of the twentieth century, and these theories have been largely based on these three characteristic features. Even so-called radical, neo-Marxist theories of development have been based on them, despite their being developed out of explicit critiques of the classic 'liberal' Rostovian stage model.

Hettne argues that a more satisfactory outcome of development efforts will come only when the models on which strategies are based are, by contrast, endogenous, normative and neo-populist. Development theory needs to move more fundamentally than it has done to derive insights into why theory has achieved so little, and why so many people in the Third World have experienced such limited development in recent decades, despite the massive efforts that have been made and the massive resources allocated to the task. The new theories that are required need to be rooted in the wants and experiences of the population involved, operating within their own cultural, social and economic behavioural assumptions. The nation state may not be the ideal unit of account, for it undermines local needs and differences. Appropriate models will be facilitated by institutions and structures that give the people 'ownership' of that development process, and an active role in defining its objectives and the means to achieving them. In most cases, and certainly in the majority of Third World countries, this will require greater recognition of geography – of regionally differentiated cultures and societies that can be more effective and cohesive building bricks.

The models of education that have been followed have certainly been mostly exogenous, borrowing heavily from a modernization perspective that assumes easy and necessary cross-cultural transferability of the institutions and content of education. The Western experience has been to the fore, not only in its structure and in the institutional forms inherited from the colonial period, but in its assumed relevance to the development of Third World countries at the end of the twentieth century. The historical experience of education in an industrializing world of manufacturing supported by low-wage labour force, that was the European and North American experience even up to the mid-twentieth century, is unlikely to be repeated in any part of the Third World, even in the NICs, and the modern economy needs to have a very different skill mix and skill levels. Just as there has been a general rejection of the diffusionist paradigm in development theory and a growing recognition of the possibilities for, or even desirability of, indigenously derived development strategies, so too in education. The possibilities for a range of locally defined strategies and modes of delivery of primary/secondary/higher, rural/urban, academic/technical certainly exist, and many cases of national experiments have been given in this book, but the possibilities for innovation seem particularly constrained on two fronts: externally by the conditionality of 'aid' programmes and multilateral and commercial loans, and internally by the demand from parents for familiar, traditional forms of schooling to which children can be sent.

The approach in education has also been positivist, assuming a universal application of 'rational' planning axioms about maximization of individual benefit and minimization of cost or effort. The evidence of the economic research on Third World education, which has been fundamental in setting the agenda of the World Bank in particular, has been widely invoked to provide a common diagnosis of problems of the education, and also to offer common solutions. There is normally formal recognition of local circumstances, but often a distinct impression of paying no more than lip service to them. The text has illustrated how needs in education and expectations of its impact are culturally variable between and within groups. The strength of local cultures of education means that the demand for schooling is not driven solely by economic motivations based on objective calculations of monetary rates of return and other assumptions that are 'rational' in the classical model. The demand is also generated by local perceptions of the contributions that schooling makes to broadening the opportunities available to children within poor communities but also more generally in the national economy.

The inherited models have also been hierarchical, bureaucratic and centralized and these have not seen great change in most countries in recent years. The years in which development planning became dominant in Third World countries, roughly 1950–80, were inevitably years of central control of that planning. Various sub-sectors, including education, were integrated into a national strategy that sought nationally defined targets and objectives. Even where education systems expanded with many more schools, teachers and enrolments, management of the larger system tended to involve deconcentration of administration down the hierarchy to provinces, districts or communes, rather than a fundamental decentralization of policy and implementation decisions. Ministries of Education remain mostly top–down in their management structures, and with central control through financial control, even if there seem to be local responsibilities for enrolment expansion at any level. In the absence of a wider political impetus for decentralization, central ministries have shown themselves to be reluctant to cede critical policy and planning powers to lower tiers of the educational hierarchy. In this respect education is little different from other areas of government activity.

Were new models of education that were fundamentally different to be introduced, however, then the contribution of education as a mechanism of change in the broader processes would be very different, and, as Hettne argues in the more general case, could be much more effective. Were they more locally ordered, more related to local cultural realities and aspirations, and more flexible in their delivery, then the education system would be more at ease with a new ideology of development that was sensitive to spatially variable needs and aspirations of a population. This would probably mean, amongst other things, much greater decentralization with stronger local powers. It would also mean much less dependence on formal acquisition of qualifications, more non-formal and adult education, more on-the-job training and a much more radical curriculum in schools in most countries. In short, a shift in emphasis from 'schooling' to 'education', but without denying the continuing importance of a formal schools system.

Developments of this type seem rather idealistic, given the pressures for continuity and conformity from above and from below. Such developments would probably result in an even greater diversity of national experiences – in the schools and in the effects of the schools on the economic and social life of the nation. The human resources would be developed in a way that allowed the system as a whole to be more responsive to local circumstances rather than to macro-economic requirements. It would cater for people's needs in their everyday lives, for the needs of farmers or industrial workers rather than those of 'manpower' categories at an aggregate scale. The effect on fertility and mortality may indeed be independent of such radical economic restructuring in any direct sense, but these components of population change would be affected by the restructuring of values in the new ideology to emphasize locally relevant skills for raising the quality of life of the population. However, there may be a further loosening of any relationships between education and migration with such changes, for with different planning and economic assumptions rural areas in the Third World could develop greater vitality and dynamism. Internationally, poor countries will gain greater self-sufficiency that creates less incentive for an external brain drain. Better provision of basic needs in such circumstances would relate schools more directly, both geographically and in terms of control and responsibility, to the people they seek to serve (Gould 1990a).

Ironically, such a radical, rather Utopian agenda need not always be at odds with the policies of 'cost sharing' and decentralization in education that are currently associated with structural adjustment packages being implemented in so many of the poorest and most indebted countries. Changes in these directions could move thinking about education and development away from approaches, notably in economics and promoted by the major international agencies and the education 'establishment', that seek a universalist view of education/development relationships. They could move the discourse towards views that are more consistent with approaches of geographers, amongst others, and so apparent at the grassroots level throughout the Third World, that recognize the importance and desirability of variability between communities in the definition and practice of 'education'. It would be a discourse more concerned to examine the contextual variables that differentiate one country from another or one region from another than to impose an 'efficient' general solution that would have equal applicability and maximum impact everywhere. Such a sea-change in thinking on development and on the role of education in development needs to build on the issues associated with many of the perspectives that have been explored in this book.

References

Abraha S, Beyene A, Dubale T, Fuller B, Holloway S, King E (1991) What factors shape girls' school performance? Evidence from Ethiopia. *International Journal of Educational Development* **11**: 107–18

Achola P P W, Kaluba H L (1989) School production units in Zambia: an evaluation of a decade of presidential experiment. *Comparative Education* **25**: 165–78

Aedo-Richmond R, Noguera I, Richmond M (1985) Changes in the Chilean educational system during eleven years of military government: 1973–84. In Brock C, Lawlor H (eds), *Education in Latin America*. Croom Helm, London, pp 163–82

Aga Khan Rural Support Programme (1988) *Fifth Annual Review, 1987*. AKRSP, Gilgit, Northern Areas, Pakistan

Altbach P G (1988) International organizations, educational policy and research: a changing balance. *Comparative Education Review* **32**: 137–42

Appleyard R (1989) Migration and development: myths and reality. *International Migration Review* **23**: 486–99

Aryeetey-Attoh S, Chatterjee L (1988) Regional inequalities in Ghana: assessment and policy issues. *Tijdschrift voor Economische en Sociale Geographie* **79**: 31–38

Awusabo-Asare K (1988) *Education and fertility in Ghana*. Unpublished Ph.D. thesis, University of Liverpool

Ayot H (1987) The harambee approach to vocational education in Kenya. In Twining J, Nisbet S, Megarry J (eds), *World Yearbook of Education: Vocational Education*. Kogan Page, London, pp 161–73

Baker V J (1989) Education for its own sake: the relevance dimension in rural areas. *Comparative Education Review* **33**: 507–26

Banya K (1991) Economic decline and the education system: the case of Sierra Leone. *Compare* **21**: 127–43

Barker D, Ferguson A G (1983) There's a gold mine in the sky faraway: rural–urban images in Kenya. *Area* **15**: 185–91

Barnett T, Blaikie P (1992) *AIDS in Africa*. Belhaven Press, London

Becker G S (1974) *Human capital: a theoretical and empirical analysis, with special reference to education*. National Bureau of Economic Research, New York

Bell M (1980) Rural–urban migration among Botswana's skilled manpower: some observations on the two-sector model. *Africa* **50**: 404–21

Bennell P (1986) Engineering skills and development: the manufacturing sector in Kenya. *Development and Change* **17**: 313–24

Berry R, Jackson R (1981) Interprovincial inequalities and decentralisation in Papua New Guinea. *Third World Planning Review* **3**: 57–77

Bhuiya A, Streatfield K (1991) Mother's education and survival of female children in a rural area of Bangladesh. *Population Studies* **45**: 253–64

Bigsten A (1984) *Education and income determination in Kenya*. Gower, Aldershot

Blaikie P (1985) *The political economy of soil erosion in developing countries*. Longman, London

Blaug M (ed.) (1968) *Economics of education 1*. Penguin Books, Harmondsworth

Bolivian Government (1990) *Demographic and health survey*. Instituto Nacional de Estadistica, La Paz

Bondi L, Matthews M H (eds) (1988) *Education and society. Studies in the politics, sociology and geography of education*. Routledge, London

Botswana Government (1977) *Education for Kagisano. Report of the National Commission on Education*. 2 Vols, Gaborone

Botti M, Carelli M D, Saliba M (1978) *Basic education in the Sahel countries*. UEI Monographs 6, UNESCO Institute for Education, Hamburg

Bourne K L, Walker G M (1991) The differential effect of mother's education on mortality of boys and girls in India. *Population Studies* **45**: 203–19

Bowman M J (1984) An integrated framework for analysis of the spread of schooling in less developed countries. *Comparative Education Review* **28**: 563–82

Brandt Commission (1980) *North–South: a programme for survival*. Pan Books, London

Bray M (1981) *Universal primary education in Nigeria. A study of Kano State*. Routledge and Kegan Paul, London

Bray M (1984) *Educational planning in a decentralized system. The Papua New Guinea experience*. University of Papua New Guinea Press, Port Moresby, and Sydney University Press, Sydney

Bray M (1986) If UPE is the answer, what is the question? A comment on the weakness in the rationale for universal primary education in less developed countries. *International Journal for Educational Development* **6**: 147–58

Bray M (1987) *Are small schools the answer? Cost-effective strategies for rural school provision*. Commonwealth Secretariat, London

Bray M, Clarke P B, Stephens D (1986) *Education and society in Africa*. Edward Arnold, London

Brock C (1985) Latin America: an educational profile. In Brock C, Lawlor H (eds), *Education in Latin America*. Croom Helm, London, pp 1–8

Brock C, Lawlor H (eds) (1985) *Education in Latin America*. Croom Helm, London

Brooke N (1992) The equalization of resources for primary education in Brazil. *International Journal for Educational Development* **12**: 37–50

Brown L A (1990) *Place, migration and development in the Third World: an alternative view with particular reference to population movements, labour market experience and regional change in Latin America*. Routledge, London

Brundtland Commission (1987) *Our common future. The World Commission on Environment and Development*. Oxford University Press, Oxford

Brydon L (1985) The Avatime family and circulation, 1900–1977. In Prothero R M, Chapman M (eds), *Circulation in Third World countries*. Routledge and Kegan Paul, London, pp 206–25

Butt M S (1984) Education and farm productivity in Pakistan. *Pakistan Journal of Applied Economics* **3**: 65–82

Caldwell J C (1968) The determinants of rural–urban migration in Ghana. *Population Studies* **22**: 361–96

Caldwell J C (1969) *African rural–urban migration: the movement to Ghana's towns*. Australian National University Press, Canberra, and Hurst, London

Caldwell J C (1979) Education as a factor in mortality decline: an examination of Nigerian data. *Population Studies* **33**: 395–413

Caldwell J C (1980) Mass education as a determinant of the timing of fertility decline. *Population and Development Review* **6**: 225–55

Caldwell J C (1986) Routes to low mortality in poor countries. *Population and Development Review* **12**: 171–200

Caldwell J C (1987) The cultural context of high fertility in Sub-Saharan Africa. *Population and Development Review* **13**: 409–37

Caldwell J C (1991) The soft underbelly of development: demographic transition in conditions of limited economic change. *Proceedings of the World Bank Annual Conference on Development Economics, 1990*. The World Bank, Washington, DC, pp 207–74

Caldwell J C, Reddy P H, Caldwell P (1985) Educational transition in rural South India. *Population and Development Review* **11**: 29–51

Carron G, Chau T N (eds) (1980a) *Regional disparities in educational development. A controversial issue*. UNESCO/ International Institute for Educational Planning, Paris

Carron G, Chau T N (eds) (1980b) *Regional disparities in educational development. Diagnosis and policies for reduction*. UNESCO/International Institute for Educational Planning, Paris

Carron G, Chau T N (1981) *Reduction of regional disparities: the role of educational planning*. UNESCO/International Institute for Educational Planning, Paris

Chau T N (ed.) (1972) *Population growth and costs of education in developing countries*. International Institute for Educational Planning, Paris

Chew S-B (1986) Human resources and growth in Singapore. In Lim C-H, Lloyd P J (eds) *Singapore: resources and growth*. Oxford University Press, Singapore, pp 119–42

Chiegwe O (1985) *Population change in southern Nigeria*. Unpublished Ph.D. thesis, University of Liverpool

Chomitz K M, Birdsall N (1991) Incentives for small families: concepts and issues. *Proceedings of the World Bank Annual Conference on Development Economics, 1990*. The World Bank, Washington, DC, pp 309–39

Cleland J G, van Ginneken J K (1988) Maternal education and child survival in developing countries: the search for pathways of influence. *Social Science and Medicine* **12**: 1357–68

Cochrane S H (1979) *Fertility and education: what do we really know?* Johns Hopkins University Press, Baltimore

Cochrane S H, Farid S M (1989) Fertility in Sub-Saharan Africa. *World Bank Discussion Papers* **43**

Colclough C (1982) The impact of primary schooling on economic development: a review of the evidence. *World Development* **10**: 167–85

Collier P, and Lal D (1986) *Labour and poverty in Kenya, 1900–1980*. Oxford University Press, Oxford

Connell J, Dasgupta B, Lashley R, Lipton M (1976) *Migration from rural areas. The evidence from village studies*. Oxford University Press, New Delhi

Conroy J D (1976) *Education, employment and migration in Papua New Guinea*. Australian National University, Canberra, Development Studies Centre, Monograph No.3

Corbridge S, Watson P D (1985) The economic value of children: a case study from rural India. *Applied Geography* **5**: 273–95

Court D (1974) Dilemmas of development: the village polytechnic movement as a shadow system of education in Kenya. In Court D and Ghai D P (eds), *Education, society and development: new perspectives from Kenya*. Oxford University Press, Nairobi, pp 219–42

Court D, Kinyanjui K (1980) Development policy and educational opportunity: the experience of Kenya and Tanzania. In Carron G and Chau T N (eds), *Regional disparities in educational development. Diagnosis and policies for reduction*. UNESCO/International Institute for Educational Planning, Paris, pp 325–409

Dale R, Esland D, MacDonald M (eds) (1976) *Schooling and capitalism: a sociological reader*. Routledge and Kegan Paul in association with the Open University, London and Henley

Dandekar H (1986) *Men to Bombay. Women at home*. Michigan Papers on South and South-East Asia 28, University of Michigan, Ann Arbor

Diab H, Wählin L (1983) The geography of education in Syria in 1882. With a translation of 'Education in Syria' by Shahin Makarius, 1883. *Geografiska Annaler* **65B**: 105–28

Dickenson J P, Clarke C G, Gould W T S, Hodgkiss A G, Prothero R M, Siddle D J, Smith C G, Thomas-Hope E M (1983) *A geography of the Third World*. Methuen, London

Doherty F J (1988) Educational provision for ethnic minority groups in Nicaragua. *Comparative Education* **24**: 103–24

Dore R (1976) *The diploma disease. Education, qualification and development*. George Allen and Unwin, London

Dorsey B J (1989) Educational development and reform in Zimbabwe. *Comparative Education Review* **33**: 40–58

El Dahab A S (1992) *Fertility and contraceptive use in Khartoum, Sudan*. Ph.D. thesis in progress, University of Liverpool

Eelens F, Speckman J D (1990) Recruitment of labor migrants for the

Middle East: the Sri Lankan case. *International Migration Review* **24**: 297–322

Ekberg N, Kinlund P, Larson U, Wählin (1985) *Geographical studies in Highland Kenya. Report from a geographical field course in Endarasha and Othaya, Nyeri District, April 1985*. University of Stockolm, Department of Human Geography, Kultur Geografiskt Seminarium 2/85

Falola J A (1989) Spatial inequalities in Nigeria's social services. In Swindell K, Baba J M, Mortimore M J (eds), *Inequality and development. Case studies from the Third World*. Macmillan, London, for The Commonwealth Foundation, pp 76–95

Farooq G M, MacKellar F L (1990) Demographic, employment and development trends: the need for integrated planning. *International Labour Review* **129**: 301–15

Ferge Z, Havasi E, Szalai J (1980) Regional disparities and educational development in Hungary. In Carron G and Chau T N (eds) *Regional disparities in educational development. A controversial issue*. UNESCO/International Institute for Educational Planning, Paris, pp 175–257

Figueroa M, Prieto A, Gutiérrez (1974) *The basic secondary school in the country: an educational innovation in Cuba*. UNESCO Press, Paris, for International Bureau of Education

Fitzgerald M (1990) Education for sustainable development. Decision making for environmental education in Ethiopia. *International Journal for Educational Development* **10**: 289–302

Foster P (1965) *Education and social change in Ghana*. Routledge and Kegan Paul, London

Foster P (1966) The vocational school fallacy in development planning. In Anderson A C, Bowman M J (eds), *Education and economic development*. Aldine Publishing Co., Chicago, pp 142–67

Foster P (1977) Education and social differentiation in less developed countries. *Comparative Education Review* **21**: 211–29

Foster P (1980) Regional disparities in educational development: some critical observations. In Carron G and Chau T N (eds) *Regional disparities in educational development. A controversial issue*. UNESCO/International Institute for Educational Planning, Paris, pp 19–48

Frank O, McNicoll G (1987) An interpretation of fertility and population policy in Kenya. *Population and Development Review* **13**: 209–43

Fredriksen B (1981) Progress towards regional targets for universal primary education: a statistical review. *International Journal of Educational Development* **1**: 1–16

Freedman R (1987) Fertility determinants. In Cleland J, Scott C (eds), *The World Fertility Survey: an assessment*. Oxford University Press for International Statistical Institute, London, pp 773–95

Freeman D B, Norcliffe G B (1984) National and regional patterns of rural non-farm employment in Kenya. *Geography* **69**: 221–33

Freire P (1972) *The pedagogy of the oppressed*. Penguin, Harmondsworth

Furter P (1980) The recent development of education: regional diversity or reduction of inequalities? In Carron G and Chau T N (eds) *Regional disparities in educational development. A controversial issue*. UNESCO/International Institute for Educational Planning, Paris, pp 49–113

Furter P (1983) *Les espaces de la formation: essais de microcomparison et de microplanification*. Presses Polytechniques Romandes, Laussane

Gerger T, Hoppe G (1982) *Education and society. The geographer's view*. Stockholm Studies in Human Geography, University of Stockholm

Ghana Government (1989) *Demographic and Health Survey*. Ghana Statistical Service, Accra

Gilbert A, Ward P H (1985) *Housing, the state and the poor. Policy and practice in three Latin American cities*. Cambridge Latin American Studies, Cambridge University Press, Cambridge

Glaser W A (1978) *The brain drain: emigration and return*. Pergamon Press, Oxford, for United Nations Institute for Training and Research

Gould W T S (1973) *Planning the location of schools: Ankole, Uganda*. International Institute for Educational Planning/ UNESCO, Paris

Gould W T S (1974) Secondary school admissions policies in Eastern Africa: some regional issues. *Comparative Education Review* **18**: 374–87

Gould W T S (1975) Movements of school children and provision of secondary schools in Uganda. In Parkin D (ed.), *Town and country in Central and Eastern Africa*, Oxford University Press, London, pp 250–61

Gould W T S (1978) *Guidelines for school location planning*. World Bank Staff Working Paper 308, Washington, DC

Gould W T S (1982a) Provision of primary schools and population redistribution, In Clarke J I, Kosinski L A (eds), *Redistribution of population in Africa*. Heinemann, London, pp 44–49

Gould W T S (1982b) Education and internal migration: a review and report with appendices. *International Journal of Educational Development* **1**: 103–11

Gould W T S (1983) The school/area controversy in migration of school leavers. *Liverpool Papers in Human Geography*, Department of Geography, University of Liverpool, No 14

Gould W T S (1985a) Migration and development in Western Kenya, 1971–82: a retrospective analysis of school leavers. *Africa* **43**: 262–85

Gould W T S (1985b) Circulation and schooling in East Africa. In Prothero R M, Chapman M (eds), *Circulation in Third World countries*. Routledge and Kegan Paul, London, pp 262–78

Gould W T S (1985c) International migration of skilled labour within Africa: a bibliographic review. *International Migration* **23**: 5–28

Gould W T S (1986) Population analysis for the planning of primary schools in the Third World. In Gould W T S, Lawton R (eds), *Planning for population change*. Croom Helm, London, pp 181–200

Gould W T S (1987) Urban bias, regional differentiation and rural–urban interaction in Kenya. *African Urban Quarterly* **2**: 122–33

Gould W T S (1988a) Urban–rural return migration in Western Province, Kenya. In I.U.S.S.P., *African Population Conference, Dakar. Conference Papers, Vol.2*, 4.1, pp 41–55

Gould W T S (1988b) Regional planning and educational planning in Sub-Saharan Africa. *Regional Planning Dialogue* **9**: 39–54

Gould W T S (1988c) Government policies and international migration of skilled workers in Sub-Saharan Africa. *Geoforum* **19**: 433–45

Gould W T S (1989) Technical education and migration in Tiriki, Western Kenya, 1902–1987. *African Affairs* **88**: 253–72

Gould W T S (1990a) Migration and basic needs in Africa. In Adepoju A (ed.), *The role of migration in African development: issues and policies for the 1990s*. Union for African Population Studies, Dakar, Senegal, pp 142–155

Gould W T S (1990b) Structural adjustment, decentralization and educational planning in Ghana. In Simon D (ed.), *Third World regional planning: a reappraisal*. Paul Chapman, London, pp 210–25

Gould W T S (1991) Missions and migration in colonial Kenya. In Dixon C, Heffernan M (eds), *Colonialism and development in the contemporary world*. Mansell, London, pp 92–105

Gould W T S (1992) Population mobility, In Gleave M B (ed.), *Tropical African development: geographical perspectives*. Longman, London, pp 284–314

Greenland J (1974) The reform of education in Burundi: enlightened theory faced with political reality. *Comparative Education* **10**: 57–63

Groth A J (1987) Third World Marxism–Leninism: the case of education. *Comparative Education* **23**: 329–44

Hackenburg R A, Magalit H F (1985) *Demographic responses to development. Sources of declining fertility in the Philippines*. Westview Press, Boulder, Co.

Hallak J (1977) *Planning the location of schools. An instrument of educational policy*. UNESCO/International Institute for Educational Planning, Paris

Hallak J, Sohrab K R, Saghafi F G, Minaie A A, Sheikhestani M S (1973) *Méthode de préparation de la carte scolaire: Le Chahrestan de Chahroud, Iran*. IIEP/UNESCO, Paris

Hallak J, Caillods F, Brjeska, Secco L (1976) *Método de preparación del mapa escolar: La región de San Ramón, Costa Rica*. IIEP/UNESCO, Paris

Hawkins J N (1983) The People's Republic of China (Mainland China). In Thomas R M, Postlethwaite N (eds), *Education in East Asia*. Pergamon, Oxford, pp 137–88

Hazlewood A, with Armitage J, Berry A, Knight J, Sabot R (1989) *Education, work and pay in East Africa*. Oxford University Press, Oxford

Helweg A W (1987) Why leave India for America? A case study approach to understanding migrant behavior. *International Migration* **25**: 165–78

Hettne B (1990) *Development theory and the three worlds*. Longman, London

Heyneman S P (1979) *Investment in Indian education: uneconomic?* World Bank Staff Working Paper 327, Washington, DC

Heyneman S P (1980a) Planning the equality of educational opportunity between regions. In Carron G, Chau T N (eds), *Regional disparities in educational development. A controversial issue*. UNESCO/International Institute for Educational Planning, Paris, pp 115–74

Heyneman S P (1980b) Differences between developed and developing countries: comment on Simmons and Alexander's 'Determinants of school achievement'. *Economic Development and Cultural Change* **28**: 403–406

Heyneman S P (1983) Improving the quality of education in developing countries. *Finance and Development* **20**: 18–21

Heyneman S P (1989) Multilevel methods for analysis of schooling effects in developing countries. *Comparative Education Review* **33**: 498–504

Heyneman S P (1990) Economic crisis and the quality of education. *International Journal for Educational Development* **10**: 115–30

Heyneman S P, White D S (eds) (1986) *The quality of education and economic development*. Proceedings of a World Bank Symposium, Washington, DC

Hill A M (1991) African demographic regimes, past and present. In Rimmer D (ed.), *Africa 30 years on*. Royal African Society in association with James Currey, London, and Heinemann, Portsmouth, NH, pp 56–72

Hinchliffe K (1989) Federal financing of education: issues and evidence. *Comparative Education Review* **33**: 437–49

Hobcraft J N, McDonald J W, Rutstein S O (1984) Socio-economic factors in infant and childhood mortality. A cross-national comparison. *Population Studies* **38**: 193–223

Holmes B (ed.) (1967) *Educational policy and the mission schools. Case studies from the British Empire*. Routledge and Kegan Paul, London

Hong S-M (1983) The Republic of Korea (South Korea). In Thomas R M, Postlethwaite N (eds), *Education in East Asia*. Pergamon, Oxford, pp 205–34

Hughes R (1991) Examining the roots of educational demand: the case of supporting rural agrarian development. *World Development* **19**: 213–24

Hunter J M (1964) Geography and development planning in elementary education in Ghana. *Bulletin of the Ghana Geographical Association* **9**: 55–64

Hyden G (1980) *Beyond ujamaa in Tanzania. Underdevelopment and an uncaptured peasantry*. Heinemann, London

Illich I (1973) *Deschooling society*. Penguin Books, Harmondsworth

International Institute for Educational Planning (1991) *Newsletter: July–September* 9

Jallade J-P (1974) *Public expenditures on education and income distribution in Colombia*. Johns Hopkins Press, Baltimore, Md.

Jamal V, Weeks J (1988) The vanishing rural–urban gap in Sub-Saharan Africa. *International Labour Review* **127**: 271–92

Jamison D T, Lau L J (1982) *Farmer education and farm efficiency*. The Johns Hopkins University Press for the World Bank, Baltimore

Jamison D T, Lockheed M E (1987) Participation in schooling; determinants and learning outcomes in Nepal. *Economic Development and Cultural Change* **35**: 279–306

Jamison D T, Moock P R (1984) Farmer education and farm efficiency in Nepal: the role of schooling, extension services and cognitive skills. *World Development* **12**: 67–86

Johnson R C (1985) Internal brain drain and development: a case study of Senegalese higher education. In Lindsay B (ed.), *African migration and national development*. Pennsylvania University Press, University Park, Pa., and London, pp 126–47

Jolly R, Colclough C (1972) African manpower plans: an evaluation. *International Labour Review* **106**: 207–64

Jones G W (1989) Sub-national population policy: the case of North Sulawezi. *Bulletin of Indonesian Economic Studies* **25**: 77–104

Keely C B (1986) Return of talent programs: rationale and evaluation criteria for programmes to evaluate a 'brain drain'. *International Migration* **24**: 179–90

Keely C B, Tran B N (1989) Remittances from labor migration: evaluations, performance and implications. *International Migration Review* **23**: 500–25

Keeves J P (1988) Cross national comparisons in educational achievement: the role of the International Association for the Evaluation of Educational Achievement. In Heyneman S P, Fägerlind T (eds) *University examinations and standardized testing. Principles, experience and policy options*. World Bank Technical Paper **78**, pp 155–66

Kelley A C, Nobbe C E (1990) Kenya at the demographic turning point? Hypotheses and a proposed research agenda. *World Bank Discussion Papers* **107**

Kelly G P (1990) Education and equality: comparative perspectives on the expansion of education and women in the post-war period. *International Journal of Educational Development* **10**: 131–42

Kenya Government (1984) *Population policy guidelines*. National Council for Population and Development, Office of the Vice-President and Ministry of Home Affairs, Nairobi

Kenya Government (1989a) *Demographic and Health Survey*. National Council for Population and Development, Nairobi

Kenya Government (1989b) *Development Plan, 1989–93*, Ministry of Planning and National Development, Nairobi

Kenya Government (1989c) *Economic Survey, 1989*. Central Bureau of Statistics, Ministry of Planning and National Development, Nairobi

Khoury E, Chemaitelly A, Wardini E, Abou-Rjaili, Zaatari M, Hajjar H (1975) *Méthode de préparation de la carte scolaire: le caza de Zahlé, Liban*. IIEP/UNESCO, Paris

King K (1971) *Pan-Africanism and education: a study of race philosophy and education in the southern states of America and East Africa*. Oxford University Press, Oxford

King K (1991) Education and training in Africa. In Rimmer D (ed.) *Africa 30 years on*. Royal African Society in association with James Currey, London, and Heinemann, Portsmouth, NH, pp 73–90

Knodel J, Wongsith M (1991) Family size and children's education in Thailand: evidence from a national sample. *Demography* **28**: 119–31

Krishan G (1989) Fertility and mortality trends in Indian states. *Geography* **74**: 53–57

Lauglo J (1992) Vocational training and 'bankers' faith in the private sector. *Comparative Education Review* **36**: 227–36

Lee K H (1988) Universal primary education: an African dilemma. *World Development* **16**: 1481–92

Leonor M D (ed.) (1985) *Unemployment, schooling and training in developing countries. Tanzania, Egypt, The Philippines and Indonesia*. Croom Helm, London, for the World Employment Programme, International Labour Office

Lewin K M, Xu H (1989) Rethinking the revolution: reflections on Chinese educational reforms. *Comparative Education* **25**: 7–17

Lieberman S S (1982) Demographic perspectives on Pakistan's development. *Population and Development Review* **9**: 85–120

Lindert P H (1983) The changing economic costs and benefits of having children. In Bulato R A, Lee R D (eds) *Determinants of fertility in developing countries. Volume I. Supply and demand for children*. Academic Press, New York and London, pp 494–516

Lipton M (1977) *Why poor people stay poor: urban bias in world development*. Temple Smith, London

Lipton M (1980) Migration from rural areas of poor countries: the impact on rural productivity and income distribution. *World Development* **8**: 1–24

Liu J M, Ong P M, Rosenstein C (1991) Dual chain migration: post-1965 Filipino immigration to the United States. *International Migration Review* **25**: 497–513

Livingstone I (1981) *Rural development, employment and incomes in Kenya: a report prepared for ILO's Jobs and Skills programme for Africa (JASPA)*. International Labour Office, Addis Ababa

Looney R L (1989) Patterns of remittances and labor migration in the Arab World. *International Migration* **27**: 563–80

Mabogunje A L (1989) *The development process: a spatial perspective*. Unwin Hyman, London

MacPhee C R, Hassan M K (1990) Some economic determinants of the Third World professional immigration to the United States: 1972–87. *World Development* **18**: 1111–18

Martin C J (1982) Education and consumption in Maragoli, Kenya: households' educational strategies. *Comparative Education* **18**: 139–55

Martin J-Y (1980) Social differentiation and regional disparities: educational development in Cameroon. In Carron G, Chau T N (eds) *Regional disparities in educational development. Diagnosis and policies for reduction*. UNESCO/International Institute for Educational Planning, Paris, pp 21–114

Masser I, Gould W T S (1975) *Interregional migration in tropical Africa*. Institute of British Geographers, Special Publication No 8, London

McNicholl G (1984) Consequences of rapid population growth: overview and assessment. *Population and Development Review* **10**: 177–240

Moock J (1973) Pragmatism and the primary school: the case of a non-rural village. *Africa* **43**: 205–15

Moock P R (1981) Education and technical efficiency in small farm production. *Economic Development and Cultural Change* **29**: 723–39

Moreland R S (1984) *Population, development and income distribution: a modelling approach*. Gower, Aldershot, and St Martin's Press, New York, for International Labour Office

Mosley P, Harrison J, Toye J (1991) *The World Bank and policy based lending (Volumes 1 and 2)*. Routledge, London

Myrdal G (1968) *Asian drama: an inquiry into the poverty of nations*. Penguin Books, Harmondsworth

Najafizadeh M, Mennerick L A (1988) Worldwide educational expansion from 1950 to 1980: the failure of the expansion of schooling in developing countries. *The Journal of the Developing Areas* **22**: 333–58

Närman A (1988) Technical secondary schools and the labour market: some results from a tracer study in Kenya. *Comparative Education* **24**: 19–35

Negash T (1990) *The crisis of Ethiopian education: some implications for nation building*. Uppsala Reports on Education, Department of Education, Uppsala, Sweden

Nyerere J *et al*. (1990) *The challenge to the South. The report of the South Commission*. Oxford University Press, Oxford

O'Connell J (1966) The political class and economic growth. *Nigerian Journal of Economic and Social Studies* **8**: 130–40

Ojo K O (1990) International migration of health manpower in Sub-Saharan Africa. *Social Science and Medicine* **31**: 361–7

Okafor S I (1989) The population factor in public service provision in Nigeria. *Applied Geography* **9**: 123–33

Ominde S H (1972) Rural economy in Western Kenya. In Ominde S H (ed.) *Studies in East African Geography and Development*. Heinemann, Nairobi, pp 207–9

Ommen T K (1989) India: 'Brain drain' or the migration of talent? *International Migration* **27**: 411–26

Peil M (1977) *Consensus and conflict in African societies*. Longman, London

Phillips D R (1990) *Health and health care in the Third World*. Longman, London

Plank D N (1987) The expansion of education: a Brazilian case study. *Comparative Education Review* **31**: 361–76

Population Reports (1982) *Population education in the schools*. Series M, **6**, Population Information Program, Johns Hopkins University, Baltimore

Population Reports (1989) *Lights! Camera! Action! Promoting Family Planning with TV, video, and film*. Series J, **38**, Population Information Program, Johns Hopkins University, Baltimore

Postlethwaite T N (1987) Comparative educational achievement research: can it be improved? *Comparative Education Review* **31**: 150–58

Postlethwaite T N, Thomas R M (eds) (1980) *Schooling in the ASEAN region*. Pergamon, Oxford

Potter R B, Unwin T (1989) *The geography of urban–rural interaction in developing countries*. Routledge, London

Preston R (1985) Popular education in Andean America: the case of Ecuador. In Brock C, Lawlor H (eds) *Education in Latin America*. Croom Helm, London, pp 92–108

Preston R (1987) Education and migration in Highland Ecuador. *Comparative Education* **23**: 191–207

Psacharopoulos G (1981) Returns to education: an updated international comparison. *Comparative Education* **17**: 321–41

Psacharopoulos G (1991) Vocational education theory, VOCED 101:

including hints for 'vocational planners'. *International Journal of Educational Development* **11**: 193–200

Psacharopoulos G, Woodall M (1985) *Education for development. An analysis of investment choices*. Oxford University Press for the World Bank, New York

Purves A C (1987) The evolution of the IEA: a memoir. *Comparative Education Review* **31**: 10–28

Rhoda R (1983) Rural development and urban migration: can we keep them down on the farm? *International Migration Review* **17**: 34–64

Richmond M (1985) Education and revolution in socialist Cuba. In Brock C, Lawlor H (eds) *Education in Latin America*. Croom Helm, London, pp 9–49

Richmond W K (1975) *Education and schooling*. Methuen, London

Riddell J B (1968) *The spatial dynamics of modernization in Sierra Leone*. Northwestern University Press, Evanston, Ill

Rocket I R H, Putnam S L (1989) Physician–nurse migration to the United States: regional and health status origins in relation to legislation and policy. *International Migration* **27**: 389–410

Rondinelli D A, Middleton J, Verspoor A M (1990) *Planning education reforms in developing countries: the contingency approach*. Duke Press Policy Studies, Durham, NC, and London

Roth G (1987) *The private provision of public services in developing countries*. Oxford University Press for the World Bank, New York

Salt J (1988) Highly skilled international migrants, careers and internal labour markets. *Geoforum* **19**: 387–99

Salt J (1989) A comparative overview of international trends and types, 1950–80. *International Migration Review* **23**: 431–56

Samoff J (1987) School expansion in Tanzania: private initiatives and public policy. *Comparative Education Review* **31**: 333–60

Schultz T W (1981) *Investing in people. The economics of population quality*. University of California Press, Berkeley

Selier F J M (1988) *Rural–urban migration in Pakistan. The case of Karachi*. Vanguard Books, Lahore

Shah N M, Al-Qudsi S S, Shah M A (1991) Asian women workers in Kuwait. *International Migration Review* **25**: 464–86

Sheffield J R (ed.) (1967) *Education, employment and rural development*. East African Publishing House, Nairobi

Shin E H, Chang K-S (1988) Peripheralization of immigrant professionals: the case of Korean physicians in the United States. *International Migration Review* **22**: 609–26

Shryock H S Jnr, Nam C B (1965) Educational selectivity of interregional migration. *Social Forces* **40**: 299–310

Simmons A, Diaz-Briquets S, Laquian A A (1977) *Social change and internal migration. A review of research findings from Africa, Asia and Latin America*. International Development Research Centre, Ottawa

Simmons O G (1988) *Perspectives on development and population growth in the Third World*. Plenum Press, New York

Simon J (1981) *The ultimate resource*. Martin Robertson, Oxford

Sinclair M E, with Lillis K (1980) *School and community in the Third World*. Croom Helm, London

Singh R D (1992) Underinvestment, low economic returns to education, and the schooling of rural children: some evidence from Brazil. *Economic Development and Cultural Change* **40**: 645–64

Skeldon R (1985) Circulation: a transition in mobility in Peru. In Prothero R M, Chapman M (eds) *Circulation in Third World countries*. Routledge and Kegan Paul, London, pp 100–20

Soja E W (1970) *The geography of modernisation in Kenya*. Syracuse University Press, Syracuse, NY

Stahl C W (1988) Manpower export and economic development: evidence from the Philippines. *International Migration* **26**: 147–70

Stock R (1985) The rise and fall of universal primary education in peripheral Northern Nigeria. *Tijdscrift voor Economische en Sociale Geografie* **76**: 277–87

Stone L (1990) Conservation and human resources: comments on four case studies from Nepal. *Mountain Research and Development* **10**: 5–6

Strudwick J, Foster P (1991) Origins and destinations in Jamaica. *International Journal of Educational Development* **11**: 149–59

Sudaprasert K, Tunsiri V and Chau T N (1980) Regional disparities in the development of education in Thailand. In Carron G, Chau T N (eds) *Regional disparities in educational development. Diagnosis and policies for reduction*. UNESCO/International Institute for Educational Planning, Paris, pp 197–323

Sutton K (1989) The role of urban bias in perpetuating rural–urban and regional disparities in the Maghreb. In Potter R B, Unwin T (eds) *The geography of urban–rural interaction in developing countries*. Routledge, London, pp 68–108

Swindell K (1970) The provision of secondary education and migration to school in Sierra Leone. *Sierra Leone Geographical Journal* **14**: 10–19

Taaffe E J, Morrill R, Gould P R (1963) Transport expansion in underdeveloped countries: a comparative analysis. *Geographical Review* **53**: 503–29

Tanzanian Government (1967) *Education for self reliance*. Dar-es-Salaam

Thabault R (1971) *Education and change in a village community. Mazières-en-Gâtine, 1848–1914*. Routledge and Kegan Paul, London

Theisen G L, Achola P P W, Boakari F M (1983) The underachievement of cross-national studies of achievement. *Comparative Education Review* **23**: 46–68

Thias H H, Carnoy M (1972) *Cost–benefit analysis in education: a case study on Kenya*. Johns Hopkins University Press, Baltimore, Md.

Thomas I D (1985) Development and population redistribution: measuring recent population redistribution in Tanzania. In Clarke J I, Khogali M, Kosinski L A (eds) *Population and development projects in Africa*. Cambridge University Press, Cambridge, pp 141–52

Thomas R M, Postlethwaite N (eds) (1983) *Education in East Asia*. Pergamon, Oxford

Thomas-Hope E M (1988) Caribbean skilled international migration and the trans-national household. *Geoforum* **19**: 423–32

Thrift N (1989) The geography of international disorder. In Johnston R J, Taylor P J (eds) *A world in crisis? Geographical perspectives*. Basil Blackwell, Oxford, pp 16–78

Timaeus I (1984) Mortality in Lesotho: a study of levels, trends and differentials, based on retrospective survey data. *World Fertility Survey Scientific Reports* **59**

Todaro M P (1976) *Internal migration in developing countries*. International Labour Office, Geneva

Toye P (1987) *Dilemmas of development. Reflections on the counter-revolution in development theory and policy*. Basil Blackwell, Oxford

Tsakok I (1982) Export of manpower from Pakistan to the Middle East, 1975–85. *World Development* **10**: 319–25

Uitto J I (1989) The Kenyan conundrum. A regional analysis of population growth and primary education in Kenya. *Meddelanden Från Lunds Universitets Geografiska Institutioner* **107**, Lund University Press, Sweden

United Nations Development Programme (1991) *Human Development Report. 1991*. Oxford University Press for UNDP, New York

UNESCO (1983) *School mapping and micro-planning in education*. UNESCO, Division of Educational Policy and Planning/IIEP, Paris

Van Raay J G T (1970) The education industry. Some geographical observations. In Mortimore M J (ed.) *Zaria and its region. A Nigerian savanna city and its environs*. Ahmadu Bello University, Department of Geography, Occasional Paper No 4, Zaria, pp 183–91

Van Rensberg P (1984) Education with production: the key to development. In Crowder M (ed.) *Education for development in Botswana*. Macmillan for Botswana Society, Gaborone, Botswana, pp 244–53

Verma M C (1985) Review of manpower forecasts in India. In Youdi R V, Hinchliffe K (eds) *Forecasting skilled manpower needs. The experience of eleven countries*. UNESCO/International Institute for Educational Planning, Paris, pp 194–210

Wagner D A (1990) Literacy assessment in the Third World: an overview and proposed schema for survey use. *Comparative Education Review* **34**: 112–38

Wählin L (1982) *Education as something new. The introduction of village schools and a study of two cohorts, 1948–1980*. The project, 'The ᶜAllan area of Jordan during one hundred years', Kulturgeografiskt Seminarium, Department of Human Geography, University of Stockholm

Watson J K P (ed.) (1982) *Education and the Third World*. Croom Helm, London

Watson T (1969) *A history of Church Missionary Society High Schools in Uganda: 1900–24: the education of a Protestant élite*. Unpublished Ph.D. thesis, University of East Africa

Werner L (1990) *Childhood mortality in Kenya*. Unpublished Ph.D. thesis, University of London

Whitehead C (1982) Education in British colonial dependencies, 1919–39: a reappraisal. In Watson J K P (ed.) *Education and the Third World*. Croom Helm, London, pp 47–60

Wood C, de Carvalho J A M (1988) *The demography of inequality in Brazil*. Cambridge University Press, Cambridge

Woodall M (1973) The economic returns to investment in women's education. *Higher Education* **2**: 275–99

Woods R I (1982) *Theoretical population geography*. Longman, London

World Bank (1974) *Education Sector Working Paper*. The World Bank, Washington, DC

World Bank (1980) *Education Sector Policy Paper*. The World Bank, Washington, DC

World Bank (1986) *Population growth and policies in Sub-Saharan Africa*. The World Bank, Washington, DC

World Bank (1987) *The Aga Khan Rural Support Program in Pakistan: an interim evaluation*. Operations Evaluation Department, The World Bank, Washington, DC

World Bank (1988) *Education in Sub-Saharan Africa. Policies for adjustment, revitalization and expansion*. World Bank Policy Paper, Washington, DC

World Bank (1990) *World Development Report, 1990. Poverty*. Washington, DC

World Bank (1991a) *World Development Report, 1991. The challenge of development*. Washington, DC

World Bank (1991b) *Vocational and technical education and training: a World Bank Policy Paper*. Human Resources Department, Education and Employment Division, Washington, DC

Wotherspoon T (1989) Immigration and the production of a teaching force: policy implications for education and labour. *International Migration* 27: 543–62

Youdi R V, Hinchliffe K (eds) (1985) *Forecasting skilled manpower needs. The experience of eleven countries*. UNESCO/International Institute for Educational Planning, Paris

Index